Karl W. Böddeker

Liquid Separations with Membranes

Karl W. Böddeker

Liquid Separations with Membranes

An Introduction to Barrier Interference

With 40 Figures and 9 Tables

 Springer

Karl Wilhelm Böddeker
Groothegen 2e
21509 Glinde
Germany

Library of Congress Control Number: 2007934042

ISBN-13   978-3-540-47451-7   Springer Berlin Heidelberg New York

Springer is a part of Springer Science+Business Media

springer.com

© Springer-Verlag Berlin Heidelberg 2008

Typesetting and Production: LE-TEX Jelonek, Schmidt & Vöckler GbR, Leipzig
Cover: WMXDesign, Heidelberg
Printed on acid-free paper   68/3180 YL – 5 4 3 2 1 0

Preface

When a fluid mixture (liquid, vapor, gas) is forced to traverse a permeable partition, such as a membrane or septum, its components are likely to move at different speed. The ensuing spread in rate and composition is a *barrier separation* effect. Unlike equilibrium separations, which depend on the thermodynamic condition of the fluid mixtures alone, barrier separations additionally are subject to specific interactions of the mixture components with the barrier. While the thermodynamics of fluid mixtures is predictable and open to adjustment, *barrier interference* adds another dimension to the repertoire of separation effects. Exploiting barrier interference is the challenge of membrane separation science and technology. This book is about the principles behind.

As *membrane processes*, barrier separations independently have acquired their peculiar identities and colorful diversity, each fondling its own tradition, each adhering to its own terminology (down to defying SI units), each earning different public attention and support. By way of illustration: Gas leakage through everything inflatable has been observed since the time the elemental gases were identified; gaseous diffusion of UF_6 through a microporous barrier, for better or worse, gave access to nuclear energy; osmotic phenomena, originally of academic interest to botanists, were engineered into providing "fresh water from the sea" by reverse osmosis; electromembrane processes, again of academic origin, now offer the vision of a clean mobility based on fuel cells; hemodialysis, starting from little known beginnings among pharmacologists, may well be viewed as the most benevolent of membrane processes; evaporation across a suitable barrier breaks azeotropes and tends to favor the higher boiling species, aqueous aromas being a case in point; microfiltration got its start at stabilizing wine through removal of microorganisms, foreshadowing the use of membranes in biotechnology.

Attempting to formally – let alone retrospectively – unify all this would be a disservice to grown variety. Yet, ignoring the fundamental kinship limits information exchange, as frequently it has in the past, and, for no good reason, burdens information dissemination as in teaching. The feature in common, and ordering principle, is the formal structure of mass transfer in barrier separation, being construed of a *driving force* (intrinsic to the fluid mixtures to be separated) and a *permeability* (summarily describing the interaction of the mixture components with the barrier). The plan of this book, accordingly, is to present the relevant thermodynamic features of fluid mixtures in contact with semipermeable barrieres, then to apply this information in deriving the working principles and design requirements of individual membrane separation processes. The membranes, by this approach, are introduced by way of the mass transport and selectivity demands which they are to meet, with due reference to the separation effects which they inspire.

The approach is made specific by examining the information needed, (a) to interpret a membrane effect (hindsight), or (b) to design a membrane separation process (foresight). In practical terms, three independent sources of information are available.

- The thermodynamic condition of the fluid mixtures to be separated, both upstream and downstream of the barrier to identify gradients. Constituting the driving force for mass transfer, this information translates into the operating conditions of the respective membrane separation process.
- The barrier (membrane) itself, its chemical nature and physical characteristic. This information comes from material science, foremost from polymer science, with additional emphasis on the art of creating thin films and microporous structures.
- The permeability of the barrier (membrane), which effectively introduces a barrier selectivity differing from equilibrium selectivity. A conglomerate of many influences, membrane permeability acquires meaning only in the context of specific applications. Independent information on polymer permeability comes from sorption and diffusion studies ("small molecule meets big molecule") originally designed to elucidate polymer structure.

This is not a compilation of expert knowledge, nor a universe of citations. Rather, an attempt is made to survey, in systematic order, the terms and concepts by which barrier separations operate, and through which practical membrane separation processes are designed.

If there is to be a motto: *The only meaningful perspective is that of context.*

Karl Wilhelm Böddeker
Technische Universität Hamburg-Harburg

Contents

Symbols and Abbreviations

Symbols

a_i	activity of i	[*concentration*]
bp.	boiling point	[°C]
c	concentration (generic)	[*various units*]
c_i	mass concentration	[kg/m^3]; [mol/m^3]
D	diffusion coefficient	[m^2/s]
d	diameter; pore diameter	[m]; [μm]
E	recovery (*Entnahme*)	[%]
G	Gibbs free energy	[J/mol]
g	acceleration of gravity	[9.81 m/s^2]
gfd	gallons per square foot & day	[*flux*]
J	mass flux (flow density)	[kg/s m^2]; [mol/s m^2]
J_p	permeance	[kg/s m^2 bar]
J_v	volume flux (practical)	[L/h m^2]
J (Joule)	energy; work (Nm)	[Ws]; [kWh]
k	mass transfer coefficient	[m/s]
L	permeability (*Leitfähigkeit*)	[*various units*]
L_p	hydraulic permeability	[*various units*]
m	molality of solute	[mol/kg]
n	number of mols	[–]
P	total pressure	[Pa]; [bar]
p	pressure	[Pa]; [bar]
p_i	partial pressure of i	[Pa]; [bar]
$p_i°$	pure component vapor pressure	[Pa]; [bar]
ppm	parts per million	[mg/L]
psi	pounds per square inch	[*pressure*]
Q	flow rate	[L/h]
R	gas constant	[8.314 J/mol K]
R (x100)	retention; rejection	[0–1] or (%)
S	sorption coefficient	[*various units*]

T	temperature	[°C]
T_g	glass transition temperature	[°C]
V	volume	[m³]; [L]; [mL]
$\overline{V}_i$	partial molar volume	[m³/mol]
W (Watt)	power (rate of work)	[J/s]
w_i	weight fraction of i	[kg/kg]
x_i	mol fraction of i (feed)	[mol/mol]
y_i	mol fraction of i (permeate)	[mol/mol]
z	distance coordinate	[m]
z	membrane thickness	[μm]

Greek

α_{ij}	separation factor	[–]
β_i	enrichment factor	[–]
γ_i	activity coefficient	[–]
δ	thickness of polarization layer	[μm]
δ	solubility parameter	[(cal/cm³)$^{0.5}$]
ε	porosity (surface or volume)	[–]
η	viscosity	[kg/s m]
μ_i	chemical potential of i	[J/mol]
π	osmotic pressure	[bar]
ρ	mass density	[kg/m³]
σ	reflection coefficient	[–]
τ	tortuosity factor	[–]

Indices

i, j	components i, j
1, 2	components 1 = solvent; 2 = solute
'	feed; feed side
"	permeate; permeate side
°	standard or reference state
b	bulk (well mixed feed)
liq	liquid
m (as index)	membrane; membrane phase
org	organic

p (as index)   permeate
v (as index)   volume
vap   vapor
w   wall (interface)
w   water

Abbreviations

ABE   acetone-butanol-ethanol
CA   cellulose acetate (generic)
CED   cohesive energy density
ED   electrodialysis
HD   hemodialysis
HF   hollow fiber
IX   ion exchange
MBR   membrane bioreactor
MD   membrane distillation
MF   microfiltration
MW   molecular weight
NF   nanofiltration
NOM   natural organic matter
PA   polyamide (generic)
PRO   pressure retarded osmosis
PV   pervaporation
RO   reverse osmosis
SDI   silt density index
SLM   supported liquid membrane
SW   seawater; spiral wound
TDS   total dissolved solids
UF   ultrafiltration
VLE   vapor-liquid equilibrium
VOC   volatile organic compound

Polymer notation

See Appendix D.

1 An Introduction to Barrier Separation

1.1 Separation is ...

Separation is the key to the uses of nature. – Gathering, harvesting, mining are elementary manifestations of selection, typifying the objective of all separation, which is added value to the product procured.

Generic categories of separation are:

- Enrichment, enhancing the proportion of a target component;
- Isolation, recovering a target product from unwanted material;
- Extraction, same when employing a liquid extractant;
- Depletion, refers to the target product in the residue of isolation;
- Purification, removing impurities from the wanted product;
- Refining, purification in specific industries or circumstances;
- Fractionation, dividing into components or component groups;
- Phase separation, parting into mutually immiscible liquid phases;
- Precipitation, rendering a solution component insoluble;
- Volume reduction, concentrating dissolved species by removal of solvent;
- Dehydration, concentrating foods and biomass by removal of water.

Membranes are instrumental in many of these.

As to technical categories of separation, King [1] lists 54 separation processes in 11 categories. Different separation processes often are applied to the same separation task, the merits of one approach then having to be assessed in comparison to others. As *membrane processes*, barrier separations add to the inventory of separation science, showing specific advantages in some applications (for example hemodialysis; azeotrope splitting; bioseparations; ultrapure water; fuel cells), while competing on equal terms with traditional

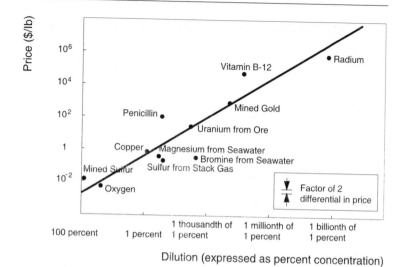

Fig. 1.1. The Sherwood plot: Selling prices of materials correlate with their degree of dilution in the initial matrix from which they are being separated. Taken from [2].

processes in many others. As a conspicuous example, membrane processes compete with distillation in water demineralization.

More often than not, separation is focused an the minority component(s) of mixtures: As wanted product to be recovered from a low-valued matrix or, conversely, as impurity to be removed to upgrade the matrix. In either mode, the expenditure to separate the minority component increases with dilution; dilution, in turn, increases with depletion. Specifically, recovery of valuable solutes from dilute liquid solutions is dominated by the cost of processing large masses of unwanted solvent. *Sherwood plots* illustrate a linear correlation between selling price of materials and their degree of dilution in the initial matrix when presented on a logarithmic scale (Fig. 1.1). Solvent removal from dilute solutions by *membrane filtration* effectively leads to solute enrichment, but just as well may serve as a means to purify the solvent.

The mechanism of separation is mass transfer. – Any mass transfer operation which produces a change in composition of a given feed mixture without permanently altering the identity of its components inherently is a separation. Any such operation yields – at least – two product mixtures which differ in composition from one another and from the original feed. If one of the products is considered the target fraction of the separation, the other, by necessity, is the original feed devoid of the target fraction. The separation effect or *selectivity* of the process is assessed by comparing the analytical composition of the two products, or by relating the composition of either one of the products to that of the original feed. The objective of separation process design usually is to render one of the products as pure as possible.

Separation is demixing. – Selective mass transfer within a multi-component system enhances the degree of order, counteracting the natural tendency to uniform mixing, and thus requires energy. According to the thermodynamics of mixtures, the *minimum energy* to isolate a pure component species from a mixture or solution is proportional to $(- ln \ x_i)$, where x_i is the mol fraction of that species in the feed mixture (Sect. 2.2.3). In terms of ordinary concentration, this proportionality is the reference coordinate of the *Sherwood plots* depicting cost of product recovery as function of initial product concentration. Conversely, the minimum energy to recover pure solvent from a given solution increases in proportion to solute concentration, affirming that the solute disturbs the thermodynamic condition of the solvent. Actual energy requirements may exceed the theoretical minimum by an order of magnitude, providing ample incentive for separation process development.

1.2   Barrier separation is ...

Barrier separation is rate controlled mass transfer. – Barrier separations rely on mass transport across *semipermeable* physical partitions, selectivity coming about by differences in *permeability* of the barrier towards the feed components resulting in the rates of mass transfer to differ. The operative distinction of rate governed versus equilibrium separation is dynamics: Mass transfer through a barrier is slowed by molecular interaction with the barrier matrix (figuratively viewed as *friction* on a molecular level), and likely is affected by encounters between the permeating species en route (loosely referred to as *coupling*); this is the essence of *barrier interference*. By comparison, mass transfer across a liquid-vapor interface (VLE

= vapor-liquid equilibrium) is considered instantaneous, and inter-action between the vaporized species is negligible.

The two generic products of barrier separation are the *permeate* (= the fraction transported through the barrier), and the *retentate* (= the fraction retained or rejected by the barrier), Fig. 1.2. Although either one may be the target fraction of the process, analysis of barrier separation is by relating the permeating fraction to the feed, thereby registering the influences of barrier interference and process conditions. Feed components present within the barrier at any time are the *permeants* (*penetrants* to some).

The term *semipermeable membrane* was introduced by van't Hoff (1887) [3], originally denoting an ideal barrier permeable to solvent (water) only while being completely impermeable to dissolved species (Sect. 3.1.2). Such a membrane would stabilize the osmotic equilibrium between a liquid solution and its own pure solvent. The contrary limit is a freely permeable, nonselective barrier yielding a permeate identical in composition to that of the feed, – in effect a throttle. Real barriers, even though selected or designed for high selectivity, are "leaky" in that, in principle, they are permeable to all species encountered. The ultimate state of a system of fluid mixtures in contact with any real membrane would be complete uniform mixing, if only one would wait long enough. There is thus no absolute barrier separation on two counts: The process is self-quenching, the energy to remove the selectively permeating species increasing with depletion; and, real membranes are leaky.

The earliest barrier on record is a section of moist pig's bladder stretched over the mouth of medicine bottles before cork stoppers came into use, hence the terms *membrane* and, in due course, *mem-*

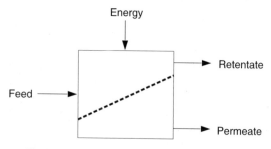

Fig. 1.2. Pictograph of a barrier separation stage.

Fig. 1.3. A sketchbook impression of Nollet's chance discovery of semipermeability: Water entering a membrane-capped vial containing "spirit of wine" creates pressure (courtesy Anne Böddeker).

brane process. Selective permeability came as a surprise to Nollet (1748) when he discovered that pig's bladder is more permeable to water than to "spirit of wine" (ethanol), resulting in a pressure phenomenon seemingly out of nowhere (Chapter 7).

1.3   Membranes, economy of size and affinity

Membranes are defined by what they do, rather than what they are. – Nature and man's ingenuity provide an abundant variety of barrier materials, both organic and inorganic, having the capacity of being permeable to individual fluids (liquids, vapors, gases), and semipermeable (selectively permeable) to fluid mixtures (Appendices D and E). The models describing membrane mass transport seek to relate *structure* and *function* of the barriers, reducing the material variety to a few phenotypes as follows.

Porous barriers, operating on size discrimination, conform to the notion of "filters": The solvent moves more or less freely, dissolved species are discriminated upon. The criterion distinguishing *membrane filtration* from ordinary (particle) filtration is solute size, smaller solutes requiring narrower pores to be retained. While gravity is all it needs to drive ordinary filtration, narrow pores require a pressure head to overcome the *hydraulic resistance* of the pore structure; the *filtration spectrum* of pressure versus solute diameter, shown in Fig. 4.2, covers the operative range of *membrane filtration*. Since specific solute species tend to be more uniform in size (respectively mass) than is met by the *pore size distribution* of most porous membrane materials, membrane characterization is in terms of rejection functions with respect to given solute size (Sect. 4.4). A pore size commensurate with a solute diameter of 0.2 μm is noteworthy in that it nominally excludes bacteria from water by *microfiltration*.

Gaseous diffusion through porous (inorganic or metallic) barriers follows a different mechanism, being governed by considerations of pore geometry versus mean free path (= pressure) of the gas or gaseous mixture components.

A survey of microporous structures is presented in Appendix E.

Homogeneous barriers (nonporous or "dense") discriminate according to relative solubilites and diffusivities of the feed components in the membrane phase. Unlike porous barriers, *solution-diffusion* type barriers rely on specific interactions of the permeants with the membrane material, its chemistry and molecular morphology. With a view at performance, more than on principle, membrane polymers (Appendix D) are assigned to one or both of the following categories,

- as *glassy* (crystalline) versus *rubbery* (elastomeric) by physical nature,
- as *hydrophilic* versus *hydrophobic* by interactive preference.

Attempting for guidance in diversity, glassy polymers generally show lower permeability and higher selectivity than rubbery ones. Liquid (aqueous-organic) separations are dominated by the sorption capacity of the membrane polymers, attended by swelling. By sorption preference, glassy polymers are hydrophilic, responding to water as being the smallest of liquid molecules at room temperature, whereas rubbery polymers tend to be organophilic (Sect. 6.2).

In practical gas separation, sorption of gases into polymers being low, the higher diffusive selectivity of glassy polymers outweighs the higher permeability of rubbery ones.

Liquid membranes function as solution-diffusion barriers, providing the very high diffusivity to permeants characteristic of the liquid state. Consequently, selective mass transport is expected to be governed by the rules of solute distribution (partition) between immiscible liquid phases in contact. *Facilitated transport* makes use of mobile carriers incorporated in the liquid membrane phase to provide species-specific selectivity.

Functionalized membranes, adding chemistry to polymer science, attempt to modify the barrier as a whole or the barrier surface to facilitate selective sorption, or else to counteract undesired membrane fouling by chemical means. A prime objective is to convey hydrophilicity to membranes used in aqueous separations, notably to reduce fouling by proteins. Another objective is resistivity towards oxidizing agents, chlorine in particular, widely employed to disinfect feed waters in water treatment.

A category of functionalized membranes of their own are charged membranes coming as anion exchangers (positive fixed charges) and cation exchangers (negative fixed charges). As "immobilized electrolytes", charged membranes are anticipated to be highly hydrophilic. When employed in *electrodialysis*, mass transport, pertaining to charged species only, is by combined action of *ionic conduction* and *Donnan exclusion* under the driving force of an electric potential.

The role of water. Water is a key component in liquid barrier separation, as is water vapor in gas separation. Not surprisingly, the presence of water within a membrane is a telltale piece of information on the nature of that membrane. With reference to the above phenotypes:

- Water in porous barriers is pore fluid. Indeed, as long as water sorption by the membrane (polymer) material itself is negligible, the difference in weight between "wet" and "dry" should equal the void space within the membrane structure (then termed *volume porosity* as against *surface porosity*, Sect. 4.2). Mass transport of solutes smaller than pore dimension is by *convection* (as in *membrane filtration*) and/or by *diffusion* within the pore fluid (as in *dialysis*). Even though mass transport is confined to the pores, the nature of the polymer matrix does matter. For example, a hydrophobic porous barrier like a microporous PTFE *(Teflon)* membrane may prevent liquid water to enter but will allow water vapor to pass (as in *membrane distillation* – and breathable textiles).

- Water absorbed (dissolved) by homogeneous polymers may be considered as a molecular solute in a polymeric solvent, causing the polymer to swell. Sorption capacity depends on the relevant interactive forces (*hydrogen bonds* and *polarity*, Sect. 6.3), but also on the "stiffness" of the polymer matrix (*glassy* versus *rubbery*) resisting polymer swelling. As an orientational aid, the dense salt rejecting layer of a composite hydrophilic membrane as employed in water desalination by *reverse osmosis* typically contains 10% of dissolved water, whereas the porous support of such a membrane may have a "porosity" (water as pore fluid) exceeding 60%.

- Charged polymers (ion exchange membranes), by both their fixed charges and mobile counter ions, provide ample ion-dipole attraction for water storage. With up to 30% of water their consistency is that of a swollen gel with restricted water mobility. However, when modeling solute mass transfer (as of ions in *electrodialysis*), ion exchange membranes are pictured as porous with the charges lining the pore walls.

1.4   Driving force, actuating barrier interference

Maxwell's demon needs help. – Next to the membranes, agents of barrier separation are the operating conditions which provide the driving force for selective mass transport

- against the inherent resistance of any mixture to demixing (this is where Maxwell's demon comes in);
- against the cohesive energy of fluid mixtures (this is where molecular interaction comes in);
- against the dynamic (transport) resistance of the barrier (this is where barrier interference comes in).

In form of the respective *gradients*, the driving force is composed of the very same variables which describe the thermodynamic condition of the fluid mixtures contacting the membrane, – *temperature*, *pressure*, and *composition*. Between them, these *intensive properties* (independent of total mass) constitute the *Gibbs free energy* or *free enthalpy* of the mixture (G). The free energy of any individual mixture component, its *partial molar free energy*, after Gibbs is named the *chemical potential* of that component species within the mixture

(μ_i). It becomes manifest as change in free energy of the mixture as the concentration of the component under consideration varies, as, for example, upon its removal in a separation process.

In actual practice, there is no need to explicitly include a temperature gradient among the driving forces since barrier separations for the most part are isothermal, usually operating at ambient (including bio-ambient) temperature. A case of exception is *membrane distillation*, which requires a thermal gradient across the porous barrier. – Likewise, an *electrochemical potential* is not included in the general treatment, electromembrane processes being confined to a class entirely of their own [12]. – On the whole, therefore, the relevant driving forces in barrier separation derive from *pressure* and *composition* of the fluid mixtures to be separated.

Pressure is the "natural" driving potential in all filtration operations, which are characterized by preferential transport of solvent (water) over solute, hence the nominal inclusion of *reverse osmosis* as "hyperfiltration" (the common expression "desalination by reverse osmosis" is misleading, "dewatering" is called forth). The upper reach of pressure encountered in membrane filtration is 100 bar (10 MPa); at this pressure ordinary liquids are incompressible, however, porous or swollen polymers are not, neither are microorganisms. – *Gas permeation* through, and *gas separation* by, homogeneous polymer membranes likewise is pressure driven, as is *gaseous diffusion* across microporous barriers.

Akin to pressure, *vapor pressure* is a driving force in barrier separation. Depending on how the vapor pressure gradient is created, the relevant membrane processes are:

- *Membrane distillation*, the only membrane process operating on a temperature gradient between liquid feed and liquid permeate. The membrane is a porous hydrophobic (water-repellent) barrier permeable to water vapor only; water transport is by evaporation into the pore space followed by re-condensation on the permeate side. – In *osmotic distillation* the vapor pressure gradient is created, not by temperature, but by a difference in solute concentration, a high solute concentration creating a vapor pressure "sink" on the permeate side, irrespective of the nature of the solute used (Sect. 2.1.1). Gentle dehydration is the usual objective of both process variants.

- *Pervaporation* is a hybrid, operating on a drastic reduction of vapor pressure (of partial pressures in case of volatile mixtures) by causing the permeants to evaporate as they emerge from the membrane. In effect, pervaporation may be viewed as nonequilibrium vacuum distillation across interacting (solution-diffusion type) barriers, usually applied to "difficult" liquid separations: Separation of narrow boiling or constant boiling (azeotropic) mixtures; separation of high boiling organics from aqueous solution (Sect. 5.3).

Composition. While pressure as driving force for mass transport conforms to intuition, *concentration gradients* do not. In fact, nature's urge to establish and maintain uniform mixing within fluid mixtures at all cost represents a powerful driving force for mass movement. It is a *virtual force*, it is the motor of *diffusion*. If, given a concentration imbalance, diffusive mixing is intercepted by a permeable barrier, mass flows will adjust themselves predictably to the permeability situation:

- A porous membrane will allow "small" solute species (including the solvent itself) to equilibrate more or less freely while retaining macromolecules. This is the operating principle of *dialysis*, hemodialysis as an example. – *Electrodialysis* is a namesake in that it, too, relocates the solute.
- With a homogeneous ("dense") membrane, if at all permeable to solvent (water), there is only one way to comply with nature's call to mitigate concentration differences: By allowing water to cross from the dilute to the concentrated side of the membrane. This is the phenomenon of *osmosis* (Sect. 3.1.2).

1.5   Dynamics of barrier separation

Mass transport is molecular motion with a directional bias. – It is slow motion, as a simple calculation will illustrate: At a throughput (*flux*) of 1000 L/d m^2 (low for ultrafiltration, high for reverse osmosis) the apparent linear velocity of mass transport within the membrane is about 4 cm/h or little more than 10^{-3} cm/s. To be sure, except for revealing a net relocation, this is no information on the actual random motion of the permeants in the membrane phase (which is a subject of *molecular modeling*).

Performance. – The formal relation between mass flux and driving force has the structure of a *generalized Ohmic law*: Flux is proportional to driving force. The coefficient of proportionality (a reciprocal resistance in the Ohmic analogy) has two meanings depending on how the driving force is introduced:

- It is a *permeability* when flux follows a *gradient* of the potential; by confining the gradient to within the membrane boundaries ("difference approximation", Sect. 2.2.2), membrane thickness becomes part of the permeability format.
- It is a *permeance* when, for a given membrane, the causality between observed flux and applied potential (as pressure or individual feed concentration) matters; it is thereby a record of performance.

$$Flux = Permeability \times Potential\ gradient$$
$$Flux = Permeance \times Potential$$

Permeability characterizes the transport capability of the barrier material itself; it thus allows for membrane material evaluation. The permeance of a given membrane (sometimes called its "productivity") is the experimentally observed flux as function of operating conditions (see Figs. 3.3; 4.4; 5.3; 5.6). If the thickness of the membrane is known, permeability and permeance correspond, permeability appearing as thickness-normalized permeance.

Barrier separations coming about through differences in transport rate of the permeants, the ratio of individual permeabilities (or permeances) suggests itself as a measure of the separation effect:

$$Selectivity\ (ij) = Permeability\ i\ (high)\ /\ Permeability\ j\ (low)$$

While this relation is formally correct, it is no recipe to estimate, much less to predict practical membrane separations, for two reasons: Individual (single component) permeabilities often are inaccessible (imagine pure salt permeability); if they are, their numerical ratio misjudges the interactions which make barrier separations interesting. It is only with true ("permanent") gases that the ratio of pure component permeabilities, individually established, quantitatively predicts the separation effect (then referred to as *ideal separation*).

Nevertheless, where accessible, single component permeability (or permeance) provides information on the intrinsic transport behavior of the barrier; pure water permeability of microporous membranes, in particular, is a key criterion in *membrane filtration*.

Concentration polarization. – The most influential effect of process dynamics on rate-governed separations by far. Referring to a gradient in composition within the feed phase next to the membrane surface, concentration polarization is a consequence of the slower permeating feed component accumulating near the solution-membrane interface as the faster permeating component moves on. As a result, the feed mixture as "seen" by the membrane differs in composition from the bulk feed, aggravating the separation task. If it is the *solvent* to permeate preferentially (as in *reverse osmosis* and all *membrane filtrations*), the *solute* being retained, concentration polarization requires conditions to be adjusted to a higher than bulk solute concentration. Conversely, if the solute or minority component permeates preferentially (as in *pervaporation* and *dialysis*), solute depletion near the membrane boundary effectively causes a lower than bulk concentration. It is to alleviate these effects that barrier separations almost always operate in the *cross flow* (tangential flow) mode, to be contrasted with *dead end* filtration.

Concentration polarization is a phenomenon to be reckoned with in liquid barrier separations. In the limit of *perfect mixing* of the feed components, as is generally the case when handling gas mixtures, the effect is irrelevant.

Whereas concentration polarization is a boundary layer effect readily rationalized, the mutual influence of permeating species on their transport behavior, referred to as *coupling*, is not easily predicted and needs case by case attention. By tendency, coupling would be expected to impair selectivity by leveling differences in mobility of the permeants, – reminiscent of the individual freedom of ions in solution being restricted by the condition of electroneutrality.

1.6   On units and dimensions

Permeability and *selectivity* are categories of performance rather than units by themselves. Reduction to practical needs is by identifying the parameters involved, both by their physical meaning and by dimension, then assigning appropriate units to the parameters identified.

It is noted that true SI units (the system dating back to 1960), besides not being universally accepted, rarely answer the needs of practical separation processing. Examples for unwieldy SI units are: Pascal (*Pa*) for pressure [replaced in this text by bar; *1 bar* = 10^5 *Pa* = *0.1 MPa*]; second (*s*) for time [in most cases replaced by hour (*h*) or day (*d*)]. Both *kg* (for "mass") and *mol* (for "amount of substance") are SI base units; yet, a *mol* of a specified substance is still a *mass* to be expressed in *kg/mol*. As an aside it is observed that industrial output is not normally reported in *mols of product*, – and if so, it would have to be *number of mols* (*n*) which, when multiplied with the respective molecular weight, is a true mass again (*kg*).

In the following some key parameters of barrier separation are discussed, using SI base units and hinting at SI derived units. It is noted that *volume*, a preeminent parameter in fluid mass transfer, is not a base unit in the SI system, although m^3 and *L* (liter) belong as SI derived.

Flux (*J*) [kg/s m^2] or [mol/s m^2]. – Flux is the quantity of permeant collected in a time (flow rate) at given membrane area, hence a *flow density* by dimension. Total flux in multiple component permeation is the sum of individual fluxes, established retrospectively by analyzing the permeate composition. Adaption to practical units, including to volume flux, is self-evident; for example, the common flux unit [L/h m^2] passes as SI derived. Dimensionally reducing a *volume flux* to a *velocity* [m^3/s m^2 → m/s], except for implementing *mass transfer coefficients* (Sect. 4.2), in most cases distracts from the physical meaning of the compound unit.

Permeability (*L*). – Permeability has many faces, all of the same dimensional configuration: Flux as function of driving force.

- The driving force for each component is a *gradient* of its chemical potential in terms of pressure or concentration (Sect. 1.4), hence the SI unit [kg m/s m^2 bar] when considering pressure-driven processes.
- Phenomenologically, permeability covers the sequence of events as a permeating mixture component makes its way from bulk feed into membrane (sorption) and thence across the membrane (diffusion), boundary layer influences and coupling effects inclusive;

it is thus a record of barrier interference. – Gas permeation is characterized by a low level of molecular interaction; individual gas permeabilities are still recorded in *Barrer units* as a semi-standard (using "cmHg" for pressure).

- Liquid permeation through porous membranes (as in membrane filtration) is described as *hydraulic permeability* (L_p); it is convective – as opposed to diffusive – volume flux (J_v) driven by a hydraulic pressure gradient [bar/m]. Pure water hydraulic permeability is one of the parameters characterizing a porous membrane. Analysis of hydraulic permeability, true to the Ohmic law analogy, is in terms of the *resistance* of the barrier to liquid transport; solute deposited on the membrane surface adds to the overall resistance (gel polarization, Sect. 4.2.2).

Permeance. – Rather than to a potential *gradient*, permeance relates the flux to the potential itself, to pressure or concentration of the permeating species. When referring to a constant pressure as driving force, permeance appears as *pressure-normalized* flux, [kg/s m² bar] in SI units. *Concentration-normalized* flux (having the dimension of a *mass transfer coefficient*), besides applying to controlled laboratory conditions, refers to separations at constant composition feed supply (seawater, for example). In batch operation, which is identical to plant operation under conditions of recovery, there is a methodical concentration dependence of flux instead (Sect. 2.2.2). – A decidedly non-SI unit of permeance is the concoction [gfd/psi], encountered in water treatment (refer to list of abbreviations).

Since sorption is prerequisite to solution-diffusion governed mass transfer, a correspondence between permeance and sorption isotherms (Sect. 2.3.1) is anticipated.

Selectivity. – Selectivity is a statement of separation performance based on a comparison of analytical compositions of feed ("bulk") and permeate. Practical needs dictate which form is used to express selectivity (Sect. 5.2 has examples). *Intrinsic selectivity* refers to the true separation capability of the barrier under undisturbed conditions, – absence of concentration polarization in particular.

Bibliography

References

[1] C. J. King: Separation Processes. Second edition, McGraw-Hill Book Company, New York, 1980.
[2] Separation & Purification, Critical Needs and Opportunities. National Research Council, National Academy Press, Washington, 1987.
[3] J. H. van't Hoff, loc. cit. Chap. 7 Ref. [9].

Text books

[4] M. C. Porter: Handbook of Industrial Membrane Technology. Noyes Publications, Park Ridge, New Jersey, 1990.
[5] W. S. Ho, K. K. Sirkar (eds.): Membrane Handbook. Van Nostrand Reinhold, New York, 1992.
[6] J. A. Howell, V. Sanchez, R. W. Field (eds.): Membranes in Bioprocessing, Theory and Applications. Chapman & Hall, London etc., 1993.
[7] R. D. Noble, S. A. Stern (eds.): Membrane Separations Technology, Principles and Applications. Elsevier, Amsterdam etc., 1995.
[8] M. Mulder: Basic Principles of Membrane Technology. Second edition, Kluwer, Dordrecht, 1996.
[9] M. Cheryan, Ultrafiltration and Microfiltration Handbook, Technomic Publishing Company, Lancaster, Pa., 1998.
[10] T. Melin, R. Rautenbach: Membranverfahren, Grundlagen der Modul- und Anlagenauslegung. Second edition, Springer-Verlag Berlin Heidelberg, 2004.
[11] R. W. Baker: Membrane Technology and Applications. Second edition, Wiley, Chichester, West Sussex, England, 2004.
[12] H. Strathmann: Ion-Exchange Membrane Separation Processes. Elsevier, Amsterdam etc., 2004.

Background reading

[13] V. Stannett, The transport of gases in synthetic polymeric membranes: An historical perspective. J. Membrane Sci. 3 (1978) 97–115.
[14] S. Loeb, The Loeb-Sourirajan membrane, how it came about. In: Synthetic Membranes (A. F. Turbak, ed.), Vol. 1, American Chemical Society, Washington, 1981.
[15] H. K. Lonsdale, The growth of membrane technology. J. Membrane Sci. 10 (1982) 81–181.
[16] E. N. Lightfoot, M. C. M. Cockrem, What are dilute solutions? Sep. Sci. Technol. 22 (1987) 165–189.

[17] E. A. Mason, From pig bladders and cracked jars to polysulfones: An historical perspective on membrane transport. J. Membrane Sci. 60 (1991) 125–145.

[18] W. J. Kolff, The beginning of the artificial kidney. Artif. Organs 17 (1993) 293–299.

[19] K. S. Spiegler, Y. M. El-Sayed: A Desalination Primer, Introductory Book for Students and Newcomers to Desalination. Balaban Desalination Publications, Rehovot, 1994.

2 The Thermodynamic Connection

2.1 Mixtures and solutions

A mixture, in dictionary parlance, is a commingling of two or more substances in varying proportion in which the components retain their individual chemical identity. Solutions, for the purpose of fluid separations, are homogeneous mixtures of solid, liquid or gaseous solutes in a liquid solvent. With uneven mixtures of two or more miscible liquids, the majority component is considered the solvent, the minority component(s) assuming the role of the solute. Even (equimolar) mixtures are exceptional.

The behavior of liquid solutions is governed by molecular interactions: Solvent-solute, solvent-solvent, and solute-solute. It is these interactions which separation has to deal with. In the following it is appropriate to distinguish between two types of solution behavior,

- the solute has no vapor pressure;
- the solute is itself volatile.

Water is the common solvent. Old alembic teaching has it that no solute will be found in the steam evolving from a boiling aqueous solution once that solute boils higher than water by upwards of 130°C. Polymer-solvent and polymer-solute interaction changes all that: Evaporation across a polymeric membrane (*pervaporation*, Sect. 5.5.2) will enrich even vanillin having a normal boiling point of 285°C from its aqueous solution, testifying to barrier interference.

2.1.1 The solute has no vapor pressure

Solutions of this type are aqueous electrolyte solutions including solutions of inorganic and higher organic acids, sugar solutions (of Pfeffer's osmotic cell fame, Sect. 3.1.1), but also true to colloidal solutions of macromolecules from proteins to microorganisms. Table 2.1 presents the range of molecular mass encountered in medical membrane use. Solubility, though ranging widely with molecular mass, always has an upper limit, solvent removal ("volume reduction") invariably leading to saturated, occasionally super-saturated solutions. True saturation would require contact with precipitated solute, a situation not normally attained with dissolved macromolecules (Sect. 4.2). If there is a solubility limit to common salts in water, it is because "free" water to effect ion hydration is no longer available: The Dead Sea, rated at 26% total dissolved solids (TDS) and lined with mineral precipitate, has the gluey consistency of glycerine.

Table 2.1. Molecular mass of nonvolatile solutes used for in-vitro clearance studies to characterize hemodialysis membranes (Sect. 4.4). After Gerner.

Solute	Molecular mass g/mol	
sodium chloride	58	small molecules
urea	60	
creatinine	113	
uric acid	168	
glucose (dextrose)	180	
sucrose	342	middle molecules
EDTA	380	
raffinose	504	
vitamin B12	1355	
inulin	5200	
$\beta2$ microglobulin	11800	large molecules
cytochrome C	13000	
hemoglobin	68000	
serum albumin	69000	

What is observed when nonvolatile solutes are dissolved in water are the *colligative properties*, which are recognized as deviations of solution properties from those of the pure solvent, namely

- lowering of vapor pressure;
- elevation of boiling point;
- lowering of freezing point;
- increase of osmotic pressure.

Strictly speaking, it is solvent properties which are affected. The colligative properties are *number effects*, depending on the molar concentration of dissolved species – ions in case of electrolytes – irrespective of their kind (and thus may be drawn upon to determine solute molecular weight). It is noted in advance that *number effects* rely on the statistical presence of the mixture components, whereas *activity effects* moreover account for molecular interactions between solute and solvent (Sect. 2.1.3).

While the first three of the colligative properties can be measured directly, osmotic pressure requires a *semipermeable membrane* to become evident. The observed effects tend to diminish with increasing molecular mass mainly because larger solute species tend to have lower solubility and thus are less "numerous". Small solutes to about MW 500 are viewed as "osmotically relevant" (Sect. 3.1.3), most common salts belonging into this category.

Considering barrier separations, solvent osmotic pressure due to the presence of nonvolatile solutes matters in *osmosis* and *reverse osmosis*. In *osmotic distillation*, a gradient in vapor pressure is generated by deliberate action of nonvolatile solutes. That action also accounts for the *salting-out* effects of organic chemistry (Sect. 5.5.2). The complex solutions of biochemical origin often contain both electrolytes and macromolecules, the separation task typically being to demineralize a macromolecular solution either by *dialysis* (as in hemodialysis) or by *ultrafiltration*.

2.1.2   The solute has vapor pressure

When two volatile liquids are mixed, the noticeable effects of solute-solvent interaction apply mutually to both. Under ideal (noninteracting) conditions, as witnessed by the absence of temperature and volume effects on mixing, the partial pressures of the two components would vary linearly with molar composition; this is *Raoult's law* for ideal solutions, and again a number effect. Real liquid mixtures deviate from Raoult's law reflecting the likes and dislikes of molecular interaction, formally accounted for by introducing a liquid phase *activity coefficient* ($\gamma \neq 1$) into Raoult's law. For ideal solutions, therefore, activity coefficients are unity.

Raoult's law, ideal solutions $\qquad p_i = x_i p^\circ = \left(1 - x_j\right) p_i^\circ$

Raoult's law, real solutions $\qquad p_i = x_i \gamma_i p_i^\circ = a_i p_i^\circ$ $\qquad$ (2.1)

Consequently $\qquad\qquad\qquad x_i$ resp. $a_i = p_i / p_i^\circ$

p_i° = pure component or saturation vapor pressure; p_i / p_i° = normalized vapor pressure. It is noted that mass action (as number effect or as activity effect) may be expressed in terms of *molar concentration* as well as *partial pressure* of the volatile solute.

The activity ($a_i = x_i \gamma_i$) replacing the analytical concentration x_i in Eq. 2.1 is an effective concentration, understood to represent the "vigor" of the component under consideration under the influence of its molecular surrounding in the mixture (Sect. 2.1.3). This influence takes one of two directions depending on the nature of the molecular interaction between solute and solvent.

Positive deviation from Raoult's law [$\gamma_i > 1$]. – On a molecular level, positive nonideality indicates repulsive interaction between dissimilar (solvent-solute) species, reflected by the solute activity coefficient to increase with dilution up to the limit of *infinite dilution* (γ^∞), in which limit the solute encounters "alien" solvent molecules only. The solvent, by then encountering "kin" only, has no reason not to behave ideally, implying $\gamma_{solvent} \approx 1$ at low solute concentration.

What is observed? Across the composition range (when going from $x_i = 0$ to $x_i = 1$) the partial pressure of any one liquid solution component increases more than proportional with concentration; so does the total vapor pressure as sum of the partial pressure contributions. Corollaries of positive solution nonideality may be a limited miscibility of the components, observed as *phase separation* or a *miscibility gap* in the phase diagram, and the occurence of *positive azeotropes* (this is where the term "positive" deviation has its origin). Positive azeotropes are constant boiling liquid mixtures of higher vapor pressure (lower boiling point) than either of the pure components. As constant boiling mixtures, azeotropes can not be separated by ordinary distillation. *Pervaporation*, a nonequilibrium membrane process, is capable of "splitting" azeotropes (Sect. 5.3).

Figure 2.1 illustrates the dependence of the activity coefficients on mixture composition of a moderately nonideal, partially immiscible aqueous-organic solution system, water-nitromethane. The system forms a positive azeotrope at 76.4 w-% nitromethane (bp. 83.6°C). It is observed that solvent (= either majority component) activity coefficients remain close to unity well into the equimolar composition range, increasing towards γ^∞ for either component with progressive dilution as shown. In the vicinity of equimolar composition, that is,

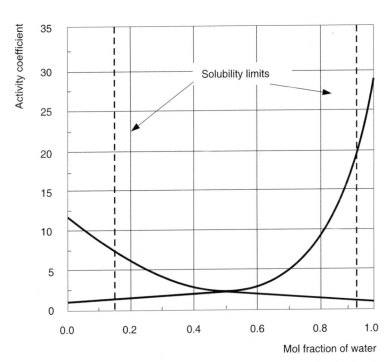

Fig. 2.1. Activity coefficients as function of mixture composition for the "positively" nonideal, partially immiscible system water-nitromethane at 50°C, indicating solubility limits [1].

even likelihood of encounter, the distinction between *concentration* and *activity* looses significance.

Of special separation concern are aqueous-organic solutions or mixtures containing sparingly soluble organic solutes, such as occur as wanted bioproducts (for example aroma compounds) or as industrial pollutants (summarily referred to as *volatile organic compounds*, VOC's). If phase-separated, a bulk aqueous phase saturated with the organic solute is in contact with a minor organic phase saturated with water, the organic phase being either distinctly separated or dispersed into "globules". At sufficiently high dissimilarity between the components, solute activity coefficient and solute solubility (= concentration) correspond inversely,

$$\gamma_{org} \approx 1 \,/\, x_{org} \qquad \text{and} \qquad x_{org} \approx 1 \,/\, \gamma_{org} \tag{2.2}$$

For a derivation consider the organic component in equilibrium across the phase boundary, implying equal organic activity in both environments ($a_{org} = x_{org}\gamma_{org}$). When viewing the organic minority phase to be itself a dilute solution (of water in organic), organic activity in that phase is unity by $x_{org} \approx 1$ and $\gamma_{org} = 1$. Equilibrium stipulates organic activity in the aqueous phase to be unity as well, hence Eq. 2.2.

The relation hints at an answer to the question "what is a dilute solution?" The answer suggesting itself at this point is: at $\gamma_{solute}^{\infty} > 100$.

Diminishing interaction between water and organic solute is evidenced by decreasing mutual solubility, resulting in the partial vapor pressures to become independent of each other. As a result, the total vapor pressure approaches the sum of pure component vapor pressures. Moreover, given a qualitative relationship between *volatility* and *solubility*, high boilers being less soluble than low boilers, the vapor phase mol fraction of a high boiling organic species in contact with its aqueous solution is expected to be close to the ratio of pure component vapor pressures, Eq. 2.3:

$$\left.\begin{array}{l} p \approx p_{water}^{o} + p_{org}^{o} \\[2ex] p_{org}^{o} \ll p_{water}^{o} \end{array}\right\} \qquad \left(x_{org}\right)_{vapor} \approx p_{org}^{o} \,/\, p_{water}^{o} \tag{2.3}$$

Steam distillation to isolate low volatile organic compounds, notably nature products such as essential oils, operates on these principles, aided by the fact that vapor pressure is an "intensive" variable: Eq. 2.3 applies independently of liquid composition, and regardless of whether the feed components are in a state of colloidal solution, microscopic dispersion, or visible phase separation, – in principle until the organic solute species is exhausted.

Negative deviation from Raoult's law [$\gamma_i < 1$]. – What is observed is a lowering of partial and total vapor pressures below those proportionate to liquid composition, qualitatively corresponding to the colligative vapor pressure lowering observed with nonvolatile solutes. On a molecular level, negative nonideality is associated with preferential interactive forces between solvent and solute, – for the most part *hydrogen bonding* and *dipole interaction* forces. As seen by the solute, these forces are most effective when no like molecules are encountered, that is, under conditions of dilution. *Solute* activity coefficients therefore decrease with dilution. Negatively nonideal liquid solutions always are miscible without limit, and may be associated with the occurence of "negative" azeotropes having a lower vapor pressure (higher boiling point) than either of the pure components. Again, *solvent* activity coefficients approach unity with decreasing *solute* concentration; again, deviation from ideal mixture behavior is smallest in the vicinity of equimolar composition.

In solution reality, *positive* deviation from Raoult's law is widespread, *negative* deviation being confined to cases of predominating solvent-solute interaction. Nearly all common aqueous-organic solution systems exhibit positive nonideality with γ_{org}^{∞} ranging from less than 2 (methanol) to a fictitious 10^{10} for nonpolar species (the range is smaller in nonaqueous systems). In separation reality, the nature of the molecular interaction bearing on the ease of separation, it is anticipated that positively nonideal liquid mixtures are easier to separate than negatively nonideal ones, including negative azeotropes. Prominent examples of negative nonideality are the aqueous solutions of simple carboxylic acids, which are notoriously difficult to separate (formic acid, above all, forming a negative azeotrope).

2.1.3   On thermodynamic activity

When an arbitrary solute species is given a chance to roam freely between two immiscible phases in contact, its analytical presence in each of the two phases is likely to differ, one environment proving more accommodating to the solute than the other. (One of the phases may well be a polymeric membrane). While this phenomenon, conforming to expectation, presents no problem to intuition, its interpretation in terms of the *thermodynamic activity* does. The condition of a distribution equilibrium across a phase boundary presupposes random movement of the solute, however, at equal rate of passage to and fro, thus maintaining the uneven distribution at any moment. The ratio of analytical ("number") concentrations is the *partition* or *distribution coefficient* of the solute (after Nernst). To account for the influence of the molecular environment on the dynamic behavior of a solute species (its "vigor"), a *solute activity* is introduced to replace *solute concentration* in such a way that, at equilibrium, activities on both sides of the phase boundary are equal. The dimensionless *activity coefficients* by which the number concentrations are modified ($\gamma \neq 1$; Eq. 2.1) in this capacity are taken to characterize solute-solvent interaction in any given solution. Since activity coefficients are typically concentration dependent (see Fig. 2.1), they are reported in the limit of infinite dilution, it being understood that the "tendency to relocate" of a solute species is highest (at $\gamma > 1$) respectively lowest (at $\gamma < 1$) when it finds itself isolated in a surrounding of solvent.

There is little practical use of assigning activity coefficients to individual inorganic ions in water. *Charge-dipole* interaction being the strongest among fluid systems, highly negative deviation from ideal solution behavior is anticipated. Moreover, though osmotically counting as individual solute species, ions are subject to the condition of electroneutrality restricting their activity. Thus when a symmetrical electrolyte dissociates, the oppositely charged ions are bound to exist at near-equal activity within their hydration shells, and *mean activity coefficients* $\gamma_{\pm} = \sqrt{\gamma_+ \cdot \gamma_-}$ are invoked to represent the activity of the salt (Appendix A).

Activity acquires a somewhat special meaning when one of the system components is largely immobile, as, for example, a membrane polymer. Mass distribution between fluid components and polymers is rationalized in terms of *sorption isotherms*, which reveal the structural identity of the polymer phase (Sect. 2.3.1).

Being associated with molecular interaction, the concept of activity looses its meaning when there are no interactive forces to speak of, as in gas mixtures under ordinary conditions of temperature and pressure. Such mixtures (which include the permeate of *pervaporation*) are described by *Dalton's law*, which states that partial pressure directly correlates with mol fraction, summing up to the total pressure of the gaseous mixture:

Dalton's law $p_i = x_i P$ $P = \sum p_i$ (2.4)

Osmotic pressures and activity coefficients of two prototype aqueous solution systems are compiled in Appendix A. One is H_2O-NaCl (a nonvolatile solute up to saturation), alluding to membranes in water demineralization; the other is H_2O-EtOH (both components volatile and miscible in all proportions), alluding to membranes in biotechnology. The data are referred to in Chapters 3 (*reverse osmosis*) and 5 (*pervaporation*).

2.2 The driving force in barrier separation

Separation is demixing, overcoming all of the tendencies which stabilize the mixture or solution, and which are mirrored by the *enthalpy of mixing*. The thermodynamic condition of a fluid mixture, its *state*, is completely described by three independent variables: Temperature, pressure, composition, – the latter in terms of mol numbers n_i. Between them, these *state variables* amount to the *Gibbs free energy* (or *free enthalpy*) of the mixture, $G(T, p, n)$. Mol numbers to express mixture composition, according to Gibbs, are state variables in that any one of them may vary independently without affecting the presence of all others (as a percentage would do). A measure of concentration in terms of mol numbers is each component's mol fraction x_i. For a binary mixture ($i = 1, 2$) these relations hold:

$$x_1 = \frac{n_1}{n_1 + n_2} \qquad x_2 = \frac{n_2}{n_1 + n_2} \qquad x_1 + x_2 = 1$$

$$\frac{n_1}{n_2} = \frac{x_1}{x_2} \qquad x_2 \approx \frac{n_2}{n_1} \quad \text{at } n_2 \ll n_1 \qquad (2.5)$$

$$dx_1 = -dx_2 \qquad dx = x \, d\ln x$$

What makes these "intensive" variables special is that they are capable of forming *gradients*, which, by their natural tendency to level, incite transport processes in the manner of a force. The same variables which define the tangible properties of a mixture also are the influence variables in separation process design, seen as change of state.

"Extensive" variables, by contrast, are volume and total free energy, but also the individual mol numbers; they depend on the "amount" present.

2.2.1   The chemical potential, no barrier

To describe a general change of state of the mixture, pictured as gradient of the Gibbs free energy (free enthalpy) over a distance coordinate z (Fig. 2.2), the total differential over the relevant variables $G(T, p, n_1, ..., n_n)$ is formed:

$$dG = \frac{\partial G}{\partial T} dT + \frac{\partial G}{\partial p} dp + \frac{\partial G}{\partial n_1} dn_1 + \frac{\partial G}{\partial n_2} dn_2 + ... \qquad (2.6)$$

This is the *Gibbs fundamental equation* for mixtures. Inspection of its terms with a view at practical barrier separations reveals:

- As a rule, liquid barrier separations operate isothermally, hence $dT = 0$ in the simplified treatment. The term thusly eliminated is the *entropy* contribution to the free energy.
- The pressure dependence of the free energy is a volume, establishing the mechanical link to the thermodynamic free energy, $\partial G / \partial p = \bar{V}$ (the bar denoting a partial molar quantity). As "pV energy", this term reiterates the dimension of the free energy as *energy* or *work*.
- The variation of the free energy of the mixture with a change in mol number of any one of its components ($\partial G / \partial n_i$) is the *partial molar free energy* of that component. Representing the contribution of mixture composition to the total free energy, Gibbs assigned the name *chemical potential* to the partial molar free energy, each component thus having its own chemical potential, μ_i. It is noted that selective removal of an individual component from a mixture or solution is the essence of separation.

$$\frac{\partial G}{\partial n_i} = \mu_i \left(T, p, x_i \right) \qquad (2.7)$$

In its reduced form the fundamental equation (2.6) now reads:

$$dG = \overline{V} dp + \sum \mu_i dn_i \qquad (2.8)$$

As in case of the total free energy, to describe a change of state of an individual component the total differential is again formed, now in terms of *partial molar quantities*:

$$d\mu_i = \overline{V}_i dp + \frac{\partial \mu_i}{\partial x_i} dx_i \qquad (2.9)$$

To formally relate the chemical potential of an individual component to mixture composition ($\partial \mu_i / \partial x_i$), Lewis introduced the concept of the *ideal mixture*, Eq. 2.10. Mixture composition is completely described by the mol fraction of the component in question; the state of reference is the chemical potential of the pure component ($\mu = \mu^o$ at $x = 1$ and $ln\, x = 0$). The Lewis concept and its implications on separation are discussed in Sect. 2.2.3.

and
$$\left. \begin{array}{l} \mu_i = \mu_i^o + RT\, ln\, x_i \\[2mm] \dfrac{d\mu_i}{dx_i} = RT \dfrac{d\, ln\, x_i}{dx_i} \end{array} \right\} \qquad (2.10)$$

When introduced into Eq. 2.9 the following elemental relation is obtained, to which reference will be made time and again when analyzing the driving force in barrier separation:

$$d\mu_i = \overline{V}_i dp + RT\, d\, ln\, x_i \qquad (2.11)$$

In terms of experimentally accessible variables, pressure and molar concentration (activity where applicable), Eq. 2.11 formulates a general (isothermal) variation of the chemical potential of an arbitrary component within an open fluid mixture. In Fig. 2.2 that variation is depicted as a linear gradient over an unconfined distance coordinate, $d\mu_i / dz$.

2.2.2 The chemical potential, barrier inclusive

While it needs potential gradients to move fluid mixture components, it takes barrier interference to sort them. As indicated in Fig. 2.2, the membrane is introduced as a partition dividing the free

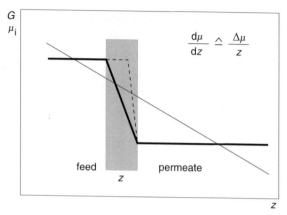

Fig. 2.2. Free energy gradient and difference approximation [2]. Within the tangible confinement of the membrane the gradient may assume different shapes as schematically indicated.

energy continuum into two realms of distinct potential, thereby transforming the potential gradient into a potential difference: The *difference approximation* [2].

$$\frac{d\mu}{dz} \approx \frac{\Delta\mu}{z} \qquad (2.12)$$

The system now consists of three phases with two phase boundaries inbetween, marking, as phase boundaries do, abrupt changes of property. The potential gradient now is confined to within the boundaries of the barrier phase of thickness z, while the potential difference to drive mass transfer across the barrier is localized in the two external phases representing *feed* and *permeate* in membrane separation. With reference to Eq. 2.11, that difference takes one of two forms, depending on whether mol fractions or partial pressures (Eq. 2.1) are used to express individual concentrations,

$$\Delta\mu_i = \overline{V}_i \Delta p + RT \ln \frac{x_i'}{x_i''}$$

$$\Delta\mu_i = \overline{V}_i \Delta p + RT \ln \frac{p_i'}{p_i''} \qquad (2.13)$$

wherein the superscripts (') and (") now indicate the feed (up-stream) and permeate (downstream) phases of the membrane system. – When likened to a concentration gradient, the shape of the potential gradient *within* the membrane reflects the swelling profile of the membrane (Fig. 2.2).

By the prevailing driving forces, practical barrier separations fall into one of two classes (not counting electromembrane processes):

- pressure driven,
- concentration (activity) driven.

Applied (external) pressure is the driving force in all *membrane filtration* processes including *reverse osmosis* (Fig. 4.2) as well as in *membrane gas separation*. Concentration (respectively partial pressure) is the driving force in *dialytic* separations as well as in *pervaporation*. Salt passage in *reverse osmosis*, as in *dialysis*, follows its own concentration gradient.

When comprehended as influence parameters in separation process design, a basic difference between the two kinds of driving force is noted: While pressure usually is maintained constant throughout the operation, the composition of the feed mixture undergoing separation varies systematically, as is the objective of selective mass transfer. *Permeance* in pressure driven membrane processes is the response of flux to operating conditions; hence the term "pressure normalized flux" (Figs. 3.3 and 4.4). *Permeance* in concentration dependent mass transfer is the response of flux to feed composition, usually with focus on the flux of the preferentially transported target species (Figs. 5.3 and 5.6).

2.2.3 Chemical potential and separation

The Lewis concept of the *ideal mixture* correlates the chemical potential (free energy) of each mixture component with its molar concentration, Eq. 2.10. The term "ideal" envisions dilute solutions of real fluids, in which both components behave ideally: The solvent as being nearly pure $(x_1 \approx 1)$; the solute as having little statistical chance to become evident $(x_2 \ll 1)$.

Addition of a solute to a solvent (which itself may be a solution) inevitably lowers the chemical potential of the solvent, as evidenced by a lowering of vapor pressure (colligative property):

$$\mu_i^o - \mu_i = \Delta\mu_i = -RT\left(ln\,x_i\right) \tag{2.14}$$

$\Delta\mu_i$ (index $i = 1$ for *solvent*) is the difference in free energy between pure solvent and solvent containing solute (note that mol fractions < 1 render $ln\,x_i$ negative). Reversing the perspective, this energy difference is the *minimum isothermal work* required to isolate pure solvent from the solution.

These are the questions of separation concern:

- How does the presence (concentration) of a solute ($x_2 = 1 - x_1$) affect the energy of separation of pure solvent from its solution (presumed dilute)?

Answer: $\Delta\mu_i$ is propotional to solute mol fraction since, for dilute solutions, $ln\,x_1 \approx -x_2$. A case in point is "water desalination" by *reverse osmosis*, a process actually involving dewatering of the saline solution (Chapter 3.2).

- How does the energy of separation of a solute species from a given solution depend on its own concentration?

Answer: $\Delta\mu_2$ is proportional to $(-ln\,x_2)$. By the arithmetic of logarithms that quantity is negative, causing $\Delta\mu_2$ to increase with dilution (respectively depletion), the limit being $ln\,x_2 = -\infty$ at $x_2 = 0$. *Sherwood diagrams* (Fig. 1.1) illustrate this correlation. Volume reduction to precede isolation would be the method of choice to recover wanted materials from dilute solutions.

As real solutions deviate from ideal mixture behavior, mol fractions no longer truly represent the concentration dependence of the free energy. Introducing the thermodynamic *activity* in place of the analytical concentration (Sect. 2.1.3) has its origin in the desire to retain the formal beauty of the Lewis concept for fluid mixtures in general. This poses the need for a convention on the condition "dilute". As implied above, aqueous solutions are considered dilute as far as the approximation $x_2 \approx -ln\,x_1$ is acceptable (x_2 for solute mol fraction). At this level, the difference between the molar volume of the solvent and its partial molar volume in solution looses meaning, too. For many practical purposes, even seawater passes as dilute solution (Sect. 3.1.3); the vapor pressure of seawater at room temperature is 1.84% below that of pure water (Spiegler).

2.3 The master flux equation

Mass transport is relocation with a directional preference under the influence of a potential gradient. One difference between fluid mixtures (such as feed and permeate phases in barrier separations) and solid solutions (such as of permeants in polymeric membranes) is that in the fluid phase all components in principle are free to move, whereas in the solid phase only the permeants have mobility, the polymer matrix remaining stationary. The segmental mobility of polymer chains, although obviously instrumental in allowing permeant relocation, has no directional bias.

Mass transport through homogeneous ("dense") membranes is by diffusion only. Mass transport within the pore space of porous barriers is by diffusion as long as the liquid phase remains stationary (*dialysis*); on applying pressure, convection superimposes diffusion (*diafiltration*).

The original statement of *diffusive mass transport* is attributed to Nernst: The rate of migration (J_i) of a species through a homogeneous fluid medium is given by the concentration of that species *in* the medium (c_i^m) times its mobility *in* the medium (u_i) under the influence of a potential gradient,

$$J_i = c_i^m u_i \, grad \, \mu_i \tag{2.15}$$

Introducing the *diffusion coefficient* to represent mobility, $u = D/RT$ (Nernst-Einstein) and applying the *difference approximation* (Eq. 2.12), a working expression for solution-diffusion mass transfer across a barrier of thickness z is obtained: The *master flux equation*.

$$J_i = \frac{c_i^m D_i}{RT \, z} \Delta \mu_i \tag{2.16}$$

By structure analogy, Eq. 2.16 is akin to an *Ohmic law* linking a current (mass or volume flux) to a potential (chemical potential difference) by way of a conductor (the permeable barrier). The permeability of the barrier is seen to be compounded of three contributing factors, unrelated in their physical nature, yet interrelated in their influence on mass transport: Sorption (c_i^m), diffusivity (D_i), membrane thickness (z). All three, properly identified as they are in the flux equation, require detailing under the circumstances of actual barrier situations, as indicated below.

By dimension, the flux in membrane operations is a *flow density*, expressed in terms of mass or volume per time and membrane area (Sect. 1.6). The *Journal of Membrane Science* lists 4 SI units and 15 "practical" units, all conforming to this dimension, to present flux.

Separation coming about by differences in the rates of mass transfer (the message of barrier interference), selectivity is defined by the *ratio of partial fluxes*. Within a self-contained membrane system that ratio reduces to the *ratio of permeabilities*,

$$J_i/J_j = \alpha_{ij} = (c_i^m/c_j^m)(D_i^m/D_j^m) \tag{2.17}$$

wherein the arrangement of terms points to the two possible mechanisms by which differentiation in mass transport according to the solution-diffusion model occurs, namely

- sorption selectivity, and (or)
- diffusion (= mobility) selectivity.

Establishing membrane selectivity (as function of feed composition and operating conditions) as a rule requires recording the separation effect on actual feed mixtures, inferring on individual (partial) flux rates from the composition of the permeate. The only exception appears to be the membrane separation of "permanent" gases, where the ratio of single gas permeabilities actually predicts the observed separation effect.

2.3.1   Sorption

Sorption (absorption) refers to the solubility of fluids (liquids or gases) in a contacting liquid or solid phase, – a polymeric membrane as a case in point. *Sorption isotherms* are a pictorial record of the equilibrium concentration of a sorbed species as function of its concentration in the external phase (external pressure in case of gases).

The simplest sorption system pictures the solubility of gases in liquids, for which *Henry's law* states that the concentration of sorbed gas is proportional to gas pressure (partial pressures in case of gaseous mixtures), $c_i = S_i p_i'$. This is the statement of a linear or "Henry-type" sorption isotherm (with S_i = sorption or solubility coefficient of component i). An example of gaseous sorption selectivity as ratio of sorption coefficients, crucial to aquatic life, is water exposed to air, the oxygen-to-nitrogen ratio in water being considerably higher than in the air above.

The approach may be generalized to apply to gas as well as to liquid sorption by polymers. In general, linear sorption is exceptional and referred to as "ideal". As illustrated schematically in Fig. 2.3, deviations from linear sorption behavior occur in both directions: *Langmuir isotherms* indicating a saturation situation, *Flory-Huggins* isotherms indicating polymer swelling (a plasticizing effect), – "ideal" sorption occuring at low sorbed concentration only.

Sorption equilibrium means equality of activity of the species under consideration between feed and membrane, $x_i' \gamma_i' = x_i^m \gamma_i^m$. Discriminating (preferential) sorption of minority solutes is the predominant mechanism of selection in liquid membrane separation, implying $x_i^m > x_i'$ and $\gamma_i^m < \gamma_i'$. The "isotherm" linking sorbed concentration with feed concentration reads

$$x_i^m = \frac{\gamma_i'}{\gamma_i^m} x_i' \qquad \text{and} \qquad S_i = \gamma_i' / \gamma_i^m \qquad (2.18)$$

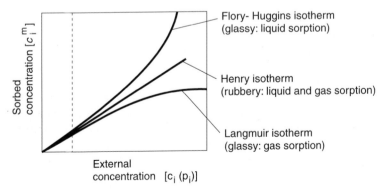

Fig. 2.3. Principal shapes of sorption isotherms (schematic). Ideal (Henry-type) sorption is found at low sorbed concentration only.

wherein the ratio of activity coefficients assumes the role of a sorption coefficient. Eq. 2.18 conveys the following information:

- Sorbed concentration – and consequently the flux – of a solute diminishes along with its concentration in the feed; hence there is a practical lower limit to recovering minority solutes, the process "slows down".
- Sorbed solute concentration increases with the degree of (positive) nonideality of the feed solution $\left(\gamma_i' > 1\right)$, resulting in improved separation selectivity (Table 5.1 as example).
- Sorbed concentration approaches feed concentration as the ratio of activity coefficients approaches unity (and vice versa); this is a statement of solvent-polymer compatibiliy otherwise known as the "like dissolves like" principle (Sect. 6.3, *solubility parameters*).

It is a trivial observation that sorption is prerequisite to permeation ($J_i = 0$ at $c_i^m = 0$, Eq. 2.16), and that flux increases with the sorption capacity of the membrane. A correlation between flux (permeance) and sorption isotherm, both functions of feed composition, is therefore expected. Examples for kinship are presented in Fig. 5.3 (Langmuir isotherms) and Fig. 5.6 (Flory-Huggins isotherms).

To the dismay of purists, liquid sorption occasionally is found to be higher than sorption from saturated vapor. The phenomenon, known as *Schroeder's paradox* [3], points to water clustering as a possible contribution to the (unwanted) salt passage through "dense" hydrophilic reverse osmosis membranes (Sect. 3.2.2).

2.3.2   Diffusivity

Diffusion contributing to membrane permeability is the mechanism of permeant transport *within* the barrier, the relevant concentration gradient being that of the sorbed species between the *internal* membrane boundaries. Diffusion is a kinetic phenomenon actuated by random thermal motion of sorbed species which are actually present, hence depending on *true local* concentration (a number effect). Influence factors, conforming to intuition, are

- size and shape of the permeants;
- the structural identity of the polymer phase; and
- permeant-polymer interaction.

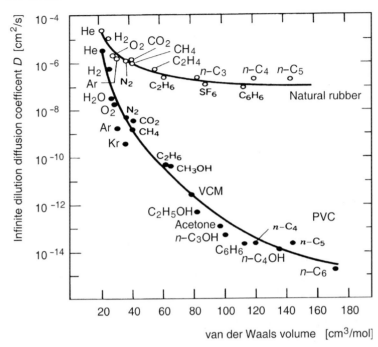

Fig. 2.4. Correlation diagram of diffusivity versus molecular size (as van der Waals volume) of low molecular weight permeants in a rubbery and a glassy polymer [4].

The illustration of Fig. 2.4, oft-quoted [4], summarizes the situation. Shown is the correlation between diffusivity and permeant size (as *van der Waals molecular volume*; alternative size indicators would have served equally well) for two polymers representing the two prototype classes of polymeric behavior described as *rubbery* and *glassy*. The considerable range of diffusion coefficients in case of the glassy ("stiff") polyvinylchloride is contrasted with the higher and less discriminating diffusivity in case of natural rubber. Diffusion selectivity (D_i / D_j) is inferred from the steepness of the slope of the tie lines between permeant pairs.

A key parameter is polymer swelling attendant to liquid sorption. As again suggested by intuition, swelling enhances permeant mobility, thereby reducing diffusion selectivity. The effect is formally accounted for by a *plasticizing parameter* χ ("Flory-Huggins interaction parameter") which renders the diffusivity of any one permeating species dependent on the local concentration of all permeants present apt to cause swelling. Local permeant concentration, in turn, is depicted as *sorption profile* across the membrane under operation conditions. In Fig. 2.2, two sorption profiles are indicated, schematically illustrating the situations of "low" and "high" swelling.

Diffusion, convection, conduction? The picture of a drop of water spreading on a piece of blotting paper is familiar. Once soaked, the wet paper will transmit water at the slightest pressure head, – gravity suffices. A stack of many wet papers will need more of a pressure; its *permeance* is reduced. This is the naive model of a hydraulic conductor of uniform water content whose *hydraulic resistance* is expected to increase with the length of the duct, eventually to the point of closure. Applying the picture to liquid barrier separations, if water is to be the preferred permeant (as in *reverse osmosis, ultrafiltration,* and *hydrophilic pervaporation*), the membrane needs to be "thin"; on the other hand, if water is to be retained (as in *organophilic pervaporation*), a "thick" membrane may be desirable.

2.3.3 Membrane thickness

How thick is "thin"? In a 1936 review of the state of the art of *ultrafiltration*, Ferry [5] ascribed the difference in behavior between then available "ultrafilter membranes" and an "ideal mechanical sieve" to the high ratio of pore length (= film thickness) to pore diameter, – which he lamented to be seldom below a thousand (alluding to μm pores in mm film).

The situation changed decisively when, around 1960, Loeb and Sourirajan discovered the "high flux" cellulose acetate membrane [6], whose structural principle was unraveled soon after by Riley: A microporous barrier integrally covered by a "dense" skin of typically $0.2\,\mu m$ (200 nm) thickness which functions as the membrane proper, – the *asymmetric membrane*. Almost immediately, the discovery elevated membranes from a laboratory tool to a technical appliance, the first aimed-for application being water demineralization by *reverse osmosis*. Although the original Loeb-Sourirajan membrane (of *cellulose diacetate*) was not yet capable to "desalt" seawater in a single pass, falling short in salt rejection, it did establish the lower limit of commercial viability of reverse osmosis in terms of permeate flux: $400\,L/d\,m^2$ (the "10 gfd criterion").

Benefiting from advances in reverse osmosis process design, both membranes and apparatus, asymmetric (effectively thin) membranes have subsequently transformed ultrafiltration (since 1965) and membrane gas separation (since 1980) into industrial separation processes as well, – strongly supported by the now legendary United States *Office of Saline Water* (OSW).

Bibliography

References

[1] S. R. Sherman, D. B. Trampe, D. M. Bush, M. Schiller, C. A. Eckert, A. J. Dallas, J. Li, P. W. Carr, Compilation and correlation of limiting activity coefficients of nonelectrolytes in water. Ind. Eng. Chem. Res. 35 (1996) 1044–1058.

[2] J. A. Wesselingh, R. Krishna: Mass Transfer. Ellis Horwood, New York etc., 1990.

[3] C. Vallieres, D. Winkelmann, D. Roizard, E. Favre, P. Scharfer, M. Kind, On Schroeder's paradox. J. Membrane Sci. 278 (2006) 357–364.

[4] R. T. Chern, W. J. Koros, H. B. Hopfenberg, V. T. Stannett, in: Material Science of Synthetic Membranes (D. Lloyd, ed.), Chap. 3, ACS Symp. Ser. 269 (1985).

[5] J. D. Ferry, Ultrafilter membranes and ultrafiltration. Chemical Reviews 18 (1936) 373–455.

[6] S. Loeb, S. Sourirajan, Sea water demineralization by means of an osmotic membrane. In: R. F. Gould (ed.): Saline Water Conversion, Adv. Chem. Series 38 (1962) 117. – See also Chap. 1, Ref. [14].

Background

[7] A. S. Michaels, H. J. Bixler, Membrane permeation: Theory and practice. In: Progress in Separation and Purification (E. S. Perry, ed.), Vol. 1, Wiley, New York, 1968, 143–186.

[8] P. Meares (ed.): Membrane Separation Processes. Elsevier, Amsterdam etc., 1976.

[9] E. L. Cussler: Diffusion. Mass Transfer in Fluid Systems. Second edition, Cambridge University Press, Cambridge, 1997.

[10] B. E. Poling, J. M. Prausnitz, J. P. O'Connell: The Properties of Gases and Liquids. Fifth edition, McGraw-Hill, New York etc., 2001.

[11] R. B. Bird, W. E. Stewart, E. N. Lightfoot: Transport Phenomena. Second edition, John Wiley & Sons, New York etc., 2002.

3 Osmosis et cetera

3.1 Osmosis

Van't Hoff's *semipermeable membrane*, postulated to advance the theory of dilute aqueous solutions, is a barrier permeable to water (solvent), while completely impermeable to dissolved solutes. It is thus a model barrier for all *membrane filtration* operations in which solutes are being retained (concentrated) by removal of solvent (Chapter 4). Osmotic effects, like all colligative properties, are confined to liquid solutions. Since nature as we know it is an aqueous system, the solvent in the following is water.

When pure water and an arbitrary aqueous solution are in contact through a semipermeable membrane at ambient pressure, pure water is "drawn" into the solution as if to dilute it: *Osmosis*. As is well known, osmosis is of utmost importance to life's functioning when comprehended as transport phenomenon on a molecular level. Living cell walls are osmotic barriers with sophisticated selectivity towards inorganic and organic solutes ("biological membranes"). The direction of osmotic water transport indicates that the solution has a lower free energy (potential) than pure water, irrespective of the nature of the solute. Specifically, it must be the activity of the solvent being lowered by influence of the solute(s), since the model barrier is presumed to communicate by way of the solvent only.

3.1.1 Osmotic investigations

When Pfeffer devised the *osmotic cell* named after him, he had plant cells in mind (1887). The original osmotic cell is an unglazed ceramic vessel of about 10 mL capacity (*A* in Fig. 3.1), to which is applied a membrane by *interfacial precipitation* as follows: The vessel, its

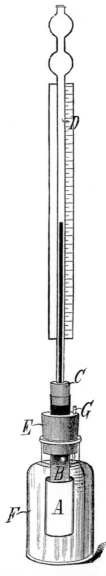

Fig. 3.1. Pfeffer's demonstration of osmotic pressure. A porous clay vessel *A*, lined with an osmotic membrane, is filled with aqueous sugar solution of known concentration; when immersed in pure water in flask *F*, readings of the osmotic pressure can be taken. From a 1909 chemistry textbook [1].

pore space soaked with water to exclude air bubbles, is briefly rinsed with a solution of copper sulfate {$CuSO_4$}, then filled with a solution of potassium ferrocyanide {$K_4[Fe(CN)_6]$} – vulgo yellow prussiate –, whereupon a precipitate of water-insoluble copper ferrocyanide {$Cu_2[Fe(CN)_6]$} forms on the inside surface. The procedure bears close resemblance to that of interfacial polymerization, by which today's composite polyamide membranes are manufactured (Sect. 6.2). By adding a minute amount of potassium prussiate to the aqueous solution being investigated, Pfeffer's membrane even has a self-mending quality to it.

The results obtained with the osmotic cell using sugar solutions in contact with pure water are as straightforward as they are puzzling: The osmotic pressure is equal to the gas pressure which would prevail if the dissolved species would fill the cell volume as an ideal gas. Thus a 0.01 molar aqueous sugar solution at room temperature exerts an osmotic pressure of 0.224 bar, increasing by 1/273 per degree of warming as postulated by Gay-Lussac's law for ideal gases.

3.1.2   The law of osmotic pressure

To quantify the effect of a solute on the free energy of the solvent, an "osmotic experiment" is visualized in which the model solution is confined in volume. The ensuing solvent influx then produces a pressure increase in the solution until a dynamic equilibrium is reached. The effect, which is in fact related to Nollet's original discovery of semipermeability (Fig. 1.3), is readily observed when ripe fruit bursts after a rain, the skin acting as semipermeable barrier enclosing the fruit tissue. The equilibrium pressure, by definition, is the osmotic pressure π of the solvent, being one of the colligative properties of liquid solutions.

At osmotic equilibrium there is no net flux, hence no discernible driving force. For $d\mu_i = 0$ at $p = \pi$ Eq. 2.11 yields ($i = 1$ for solvent; $i = 2$ for solute):

$$V_i \pi = -RT \ln x_i \quad \text{and} \quad \pi = -\frac{RT}{V_i} \ln x_i \qquad (3.1)$$

For dilute solutions ($x_2 \ll x_1$) the mol fraction of solute responsible for the osmotic pressure of the solvent is approximated by $x_2 = -\ln x_1$ (Sect. 2.2.3), which, when introduced into Eq. 3.1, gives

$$ \pi = \frac{RT}{V_1} x_2 \qquad or \qquad \pi V_1 = RT\, x_2 \qquad (3.2) $$

for the osmotic pressure of the solvent as function of the molar concentration of a solute, irrespective of the nature of the solute ("number effect").

Equation 3.2 is known as van't Hoff's limiting law of osmotic pressure (1886), the attribute "limiting" alluding to the concept of an *ideally semipermeable* membrane in contact with a *dilute* solution. Intrigued by Pfeffer's osmotic cell results, the semblance of van't Hoff's law with the ideal gas law gave rise to the notion that osmotic pressure has the same kinetic origin as an ordinary gas pressure: Momentum transfer of thermally agitated species bouncing against the wall of their containment. On closer inspection, this is a speculative notion at best. In fact, as follows from a surface tension argument, electrolytes in aqueous solution even tend to withdraw from the phase boundary. (A proposition to recover desalted water by *surface skimming* of seawater, based on this argument, did not materialize, however.)

The relevant influences on the free energy of the solvent of liquid solutions are now seen to be reduced to two formally related, yet fundamentally counteracting variables: External pressure (the pV term) raises the potential of the solvent (of virtually anything), while the presence of a solute (the πV term) lowers it, Fig. 3.2. When, in an osmotic experiment, external pressure outweighs the effect of osmotic pressure, the direction of solvent flow is reversed: *Reverse osmosis*. Solvent flow in reverse osmosis thus relies on the increment of external pressure over and above the osmotic pressure of the feed solution.

While practical reverse osmosis is concerned with solutions of solutes having no vapor pressure (such as H_2O-NaCl, Appendix A upper), there is no difference in principle when turning to mixtures both components having vapor pressure (such as H_2O-EtOH, Appendix A lower). The additional aspect is that by then both mixture components mutually exert solute effects resulting in osmotic pres-

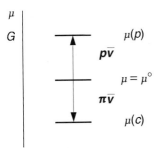

Fig. 3.2. Hydrostatic pressure (*p*) and osmotic pressure (*π*) counteracting in their effect on solvent free energy as *pV energy* (mechanical) versus *πV energy* (chemical).

sure (colligative properties) for each. Referring to Raoult's law, Eq. 2.1, a rendition of osmotic pressure in terms of vapor pressure is

$$\pi(i) = \frac{RT}{V_i} \, ln\left(\frac{P_i^{\circ}}{P_i}\right) \tag{3.3}$$

where *i* now is solvent or solute depending on proportion. Given sufficient mutual miscibility, osmotic pressures of aqueous-organic solutions may reach extreme values, see Appendix A.

3.1.3   Osmotic pressure illustrated

Nature provides two semi-standards of osmotic pressure, the pre-eminence of which remains a matter of speculation.

- One is the equivalent salinity of the **body fluids** of warm-blooded mammals, which is osmotically matched by an *isotonic solution* of 0.9 w-% NaCl at about 7.5 bar osmotic pressure (referred to as "saline" in Fig. 4.4). A physiological blood infusion (*Ringer solution*) is prepared, for example, by dissolving NaCl (8.0 g); KCl (0.2 g); CaCl$_2$ (0.2 g); MgCl$_2$ (0.1 g); NaHCO$_3$ (1.0 g); NaH$_2$PO$_4$ (0.05 g); glucose (1.0 g) in 1 L water. Whole blood plasma, in addition to "salts", contains macromolecular species (proteins). Therapeutic protein substitution is by hydrophilic (soluble) polymers in concentration to match the incremental osmotic pressure of high molecular weight blood components. A telltale example is polyvinylpyrrolidone (PVP) which, it is noted, plays a key role in continuing efforts to convey hydrophilicity to synthetic membranes, including those employed in hemodialysis (Sect. 6.2).

- The other is *seawater*, which covers three quarters of the globe at remarkably uniform mineral content, mineral composition and density (Sect. 3.3.1). A concentration of 3.45 w-% total dissolved solids (34500 ppm TDS of "sea salt") associated with 25.5 bar osmotic pressure sometimes is taken as *standard seawater*. Computed in terms of the ionic concentrations of the major inorganic constituents, seawater has a *salinity* of approximately 1 mol/L. Local variations in the salinity of real seawater do occur, however, and have a significant effect on the performance (= economics) of water desalination: The majority of seawater desalination plants operate from confined seawater bodies such as the Persian Gulf and the Red Sea having higher than open-ocean salinity.

More osmotic pressure. – For the following observations the data tabulated in Appendix A serve as illustration. According to Eq. 3.2, solvent (water) osmotic pressure increases with solute concentration, a linear dependence being observed well into an intermediate composition range for both NaCl and EtOH as solutes. The same is true for water as solute in ethanolic solution, the lower slope indicating a difference in solute activity. At comparable mol fraction of solute (for example, $x_2 = 0.01$), osmotic pressure in case of NaCl is twice that observed with EtOH (26 versus 13 bar), electrolytic dissociation doubling the number of "osmotically relevant" species with the salt.

Turning to the completely miscible system H_2O-EtOH, it is observed that the osmotic pressure of either component increases with progressive dilution, Eq. 3.4. To verify the data would require two separate "osmotic experiments" using membranes of contrasting permeability description,

- one a membrane preferentially permeable to water to ascertain the effect of an organic solute (a *hydrophilic* membrane);
- the other preferentially permeable to organic solvents to ascertain the effect of water as solute (an *organophilic* or *hydrophobic* membrane).

Reverse osmosis is imaginable in either direction, provided semi-permeable membranes as described are available. The principal and practical limitation is the osmotic pressure which needs to be overcome (referred to as *osmotic pressure limitation*), as again revealed by the data of Appendix A. For example, the minimum pressure to dehydrate wine (11.9% EtOH) by an ideally hydrophilic barrier is 64 bar; conversely, at least 860 bar is needed to remove pure ethanol from wine through an organophilic barrier. Removing water from the ethanol-water azeotrope (4 w-% H_2O at 78°C) by reverse osmosis would require pressures in excess of 2000 bar; the other way around is meaningless. – There is in fact a membrane process capable of splitting azeotropic mixtures, which relies on a drastic reduction of the activity of the permeate by causing it to evaporate: *Pervaporation*.

As examples encountered in food processing, Table 3.1 lists osmotic pressures of various juices. Membranes have the capacity to concentrate bioorganic solutions under "mild" (low temperature) conditions, retaining aroma compounds; however, as the figures indicate, dehydration by reverse osmosis faces the osmotic pressure limitation. A suitable dehydration process independent of this limitation is *membrane distillation* and its counterpart, *osmotic distillation* (Sect. 1.4).

Table 3.1. Osmotic pressure of fruit juices and milk [2]

Juice	Concentration (degree Brix)	Osmotic pressure (bar)
Sugar beet juice	20	34.5
Cane sugar juice	44	69
Tomato juice	33	69
Lemon juice	10	15
Lemon juice	45	103
Orange juice	10–12	17–20
Orange juice	42	103
Orange juice	60	207
Milk (for comparison)	~ 12% TS	6

Brix: An industrial measure related to density (10 Brix ≈ 1 w-% sugar at 20°C).
TS: Total solids (lactose and salts).

3.2   Reverse osmosis

By generic category, reverse osmosis is volume reduction through selective removal of solvent, the driving force being an external pressure over-compensating the osmotic pressure of the feed solution. Because of the osmotic pressure limitation, reverse osmosis separations focus on solvent recovery (as permeate) more than on solute enrichment (in the retentate).

Practical reverse osmosis (RO) aims at recovering demineralized water from natural saline solutions – seawater and brackish waters – by pressure-driven permeation through hydrophilic polymer membranes (slogan "fresh water from the sea"). Central to design and operation of the process is the osmotic pressure of the feed solution, the pressure of reference being the osmotic pressure of "standard seawater" of 25.5 bar at 25°C.

3.2.1   Solvent flux and solute rejection

Real membranes are "leaky", never completely excluding unwanted feed components. Modeling reverse osmosis desalination therefore requires to consider both water transport and salt passage. The *Merten model* of mass transport in reverse osmosis [3] is based on these premises:

- Mass transport is by a solution-diffusion mechanism (Sect. 1.3.2);
- there is no coupling between water and salt transported, permitting the master flux equation (Eq. 2.16) to be applied independently to each;
- the membrane is highly salt rejecting, implying the difference in salinity between feed and permeate (Δc_2) to be high; and
- the model holds as far as the approximation "dilute" applies (Sect. 2.2.3), implying ideal solution behavior for the solvent, and negligible difference between the molar volume of water and its partial molar volume in solution $\left(\overline{V}_1 \approx V_1\right)$.

Solute concentration (c_2 in the following) is summarily given as w-% or *ppm* (= mg/L) of "salt", linked in practice to electrical conductivity. Justification comes from the fact that, even though natural saline waters (seawater) contain a multitude of inorganic salts, remaining salt in the permeate of reverse osmosis (and by implication within the membrane) predominantly is NaCl; the permeate is "soft" (Appendix B).

Solvent (water, index 1). – The driving force for solvent flow is obtained by substituting the concentration term in Eq. 2.10 by the corresponding osmotic pressure (Eq. 3.1) to obtain $\Delta\mu_1 = V_1(\Delta p - \Delta\pi)$. With this rendition of the solvent driving force the flux equation, Eq. 2.16, becomes

$$J_1 = \frac{c_1 D_1}{RT\, z}\; V_1\left(\Delta p - \Delta\pi\right) \tag{3.4}$$

To picture the simplest situation, permeate pressure is ambient and permeate solute concentration is low, in which case both pressure terms in Eq. 3.4 refer to the condition of the feed.

Solute ("salt", index 2). – Since the solute has no vapor pressure, there is no osmotic pressure equivalent to concentration. Instead, the total differential of its chemical potential with respect to the variables pressure and concentration is again formed, $\mu_2(p;\, c_2)$. Observing that $\partial\mu_2/\partial p = \overline{V_2}$ (the partial molar volume of the solute), and noting that $d\ln c/dc = 1/c$, the following relation for the driving potential for the solute is obtained:

$$\Delta\mu_2 = \overline{V_2}\,\Delta p + \frac{RT}{c_2}\,\Delta c_2 \tag{3.5}$$

Adaptation to the premises of the Merten model is by observing that the effect of pressure on salt passage is small compared to the influence of the concentration gradient between feed and permeate, Δc_2. Furthermore, the concentration c_i^m of the master flux equation (Eq. 2.16) is that of the permeant *within* the diffusion barrier, while the salinity which determines the driving force is that of the *external* feed solution, c_2'. The distribution coefficient relating the two, $c_2^m = Kc_2'$, makes salt uptake by the membrane a function of the salinity of the feed solution contacting it. All things considered, the flux equation for salt passage in reverse osmosis reduces to

$$J_2 = \frac{KD}{z}\,\Delta c_2 \tag{3.6}$$

Salt rejection. – The selectivity of reverse osmosis desalination is expressed as *salt rejection R* in terms of the analytical (bulk) solute concentrations of feed (c_2') and permeate (c_2'') as

$$R = \frac{c_2' - c_2''}{c_2'} = 1 - \frac{c_2''}{c_2'} \qquad R \times 100 = [\%] \qquad (3.7)$$

For example, desalting standard seawater (34500 ppm TDS) to potable water (< 500 ppm TDS) nominally would require a salt rejection of 98.6%. Under dynamic process conditions a higher than nominal rejection is needed (Sect. 3.2.3). As the model implies, rejection requirements depend on the desalting task: Less severe on lower salinity feed waters; more stringent when a high degree of demineralization is needed.

In place of percentage rejection, the *salt passage (100 – R)* is used in industry. In ultrafiltration a *sieving coefficient* is used, similarly describing selectivity in terms of solute whereabouts (Sect. 4.4).

3.2.2 Model implications

Although based on low-recovery laboratory conditions (which yield "intrinsic" membrane performance data), the Merten model represents actual reverse osmosis performance using available membranes to a fair degree. The principal model predictions, illustrated schematically in Fig. 3.3, are as follows.

Mass transport. – Water flux (in $kg/d\,m^2$), commencing at the osmotic pressure of the feed solution, increases linearly with external pressure, Eq. 3.4. Salt flux (in $g/d\,m^2$; note a factor of 10^3 between fluxes) is governed by the difference in salinity between feed and permeate, Eq. 3.6, and is *not* affected by pressure. Salt transport in reverse osmosis thus is a *dialysis* phenomenon, the level of salt passage depending on the nature of the membrane polymer. The resultant solute rejection R is a function of pressure simply because a higher rate of solvent flux relative to a constant salt flux results in enhanced permeate dilution. As a consequence, to achieve the salt reduction desired, applied reverse osmosis operates at pressures of typically twice the osmotic pressure of the feed solution, $p \geq 2\pi$. This condition contributes to the osmotic pressure limitation of practical reverse osmosis towards high feed salinities.

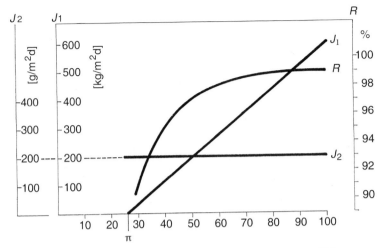

Fig. 3.3. Water flux (J_1), salt transport (J_2) and resultant salt rejection (R) in reverse osmosis desalination of seawater in the approximation of the Merten model.

Adjustable parameters of membrane performance, according to the model, are the water content of the membrane (c_i^m in Eq. 3.4) and its thickness (z). For favorable selectivity, water sorption should be high, salt sorption (as distribution coefficient, K) low. Commonly used hydrophilic membrane polymers, in the main *aromatic poly-amides* and *cellulose esters,* have sorption capacities for liquid water of the order of 10% ("primary water"), equivalent to an increase in volume (swelling) of the same order. At higher water uptake than allowable to sustain a solution-diffusion mechanism of mass transport, sorbed water tends to aggregate into liquid domains (droplets), causing a breakdown of salt rejection. There is little variance of water content across a swollen reverse osmosis membrane in operation, providing an isotropic passageway for diffusive mass transport across (Sect. 5.4.2). Water is "faster" than salt by orders of magnitude.

The membrane. – The permeation rates of solvent and solute are both inversely related to membrane thickness, implying that the selectivity – as ratio of the two rates – is independent of membrane thickness. High solvent flux (*permeance*) at unimpaired selectivity thus demands the membranes to be thin. It was the discovery of the *asymmetrically structured* ("skinned") cellulose acetate membrane by Loeb and Sourirajan (1960) which paved the way to *high flux membranes* (Sect. 2.3.3), (and on which the Merten model is based).

Today's *composite membranes* are asymmetric as well, the active "dense" skin being produced by interfacial polymerization onto a porous support. A semi-standard in seawater reverse osmosis is a composite membrane of an aromatic polyamide ($< 0.2\,\mu m$) on a polysulfone microporous support ($40\,\mu m$), mechanically stabilized by a polyester fabric, Fig. 3.4. An electron micrograph of the skin section of such a membrane is shown in Appendix E, Fig. E.6.

The development of seawater reverse osmosis membranes in terms of "intrinsic" performance is summarized by the data of Table 3.2 [5].

Historically, the original asymmetric cellulose acetate membrane of Loeb-Sourirajan attained an intrinsic flux of $400\,L/d\,m^2$ ($= 10\,gfd$) at salt rejection of 96%. To this day, that flux marks the lower limit of economic viability in reverse osmosis desalination (the "10 gfd criterion").

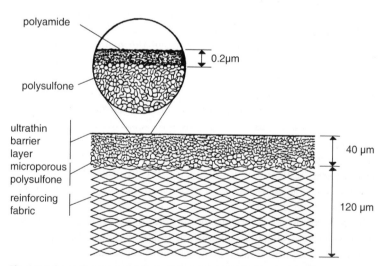

Fig. 3.4. Schematic representation of the make-up of a thin-film composite membrane as employed in seawater desalination (FT-30; FilmTec/Dow). The active layer is an aromatic polyamide produced by in situ polymerization [4].

Table 3.2. Representative "intrinsic" performance record of seawater reverse osmosis membranes. CA = cellulose acetate; PA = polyamide [5].

	Year	Flux $L/d\,m^2$	Salt Rejection %
CA	1978	650	98.9
PA	1986	1300	99.4
	1995	1300	99.7
	2004	1500	99.8

Test conditions: 32000 ppm NaCl; 55 bar; 25°C; recovery < 10%

Two comments are in order:

- Membrane performance under actual production conditions necessarily is below intrinsic: Dynamic reverse osmosis, Sect. 3.2.3.
- A major improvement of membrane performance along present lines of development seems unlikely; improvement of overall process economy therefore focuses on auxiliary technology: Energy recovery (Sect. 3.2.4) and feed water conditioning (Sect. 3.2.5).

3.2.3 Dynamic reverse osmosis

With reference to the flow scheme of Fig. 3.5, reverse osmosis in operation is described as follows. The feed solution of analytical (bulk) concentration $c_o = c_2'$ and corresponding osmotic pressure π_o is pumped to the membrane stage (a single "module" or an alignment of modules) at *feed flow rate* Q_o (L/h) and *feed pressure* (operating pressure) p_o consistent with the osmotic pressure of the feed, $p_o \geq 2\pi$. System pressure is established by a back-pressure regulator (adjustable valve) in the *retentate* (reject) stream, Q_s. En route through the membrane stage of specified membrane area, demineralized permeate (the low salinity product water) is withdrawn from the feed stream at an integral flow rate Q_p, resulting in a gradual build-up of solute concentration from c_o (inlet) to c_s (outlet), along with a decrease in volume feed flow commensurate with the overall flow balance, $Q_o = Q_s + Q_p$. Membrane performance (permeance), being a function of local feed concentration (increasing) and local feed pressure (decreasing), declines systematically from inlet to outlet of the membrane stage.

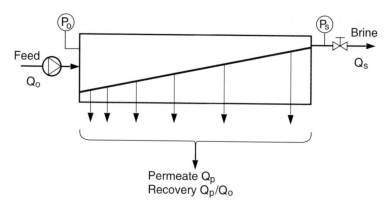

Fig. 3.5. Dynamic reverse osmosis: Flow pattern and process parameters schematically representing a single membrane stage in operation. Q = flow rate (L/h); P = hydrostatic pressure; E = recovery.

Process design (including membrane selection) depends on the osmotic pressure of the feed solution, being a function of solute concentration, by

- determining the operating (= inlet) pressure of the process;
- limiting the permeate (product) recovery from a given feed stream;
- limiting the degree of solute concentration build-up in the retentate; and
- determining the energy requirements of the separation process.

Process analysis is by inspecting the two – interrelated – mechanisms by which solute concentration builds up: *Permeate recovery* and *concentration polarization*.

Recovery. – Permeate recovery (also termed *recovery ratio* or *conversion*) represents the yield of the separation process, that is, its purpose. Recovery is the ratio of permeate flow rate to feed flow rate, usually given as percentage of the feed volume,

$$E = Q_p / Q_o \qquad E \times 100 = [\%] \qquad (3.8)$$

Solute concentration build-up as a result of permeate recovery reduces the available pressure head for solvent flux (Eq. 3.4) and increases the permeation rate of solute (Eq. 3.6), both effects, even though intended, predictably to the disadvantage of process performance.

There are two modes of reverse osmosis processing:

- In *solvent recovery*, operation is out of an infinite feed reservoir (such as a seawater supply), implying fixed inlet conditions. Plant (stage) recovery is a function of feed flow rate and membrane area provided. Within every stage (such as schematically depicted in Fig. 3.5) conditions correspond to those of *volume reduction*. Solvent recovery is equivalent to once-through batch processing, – as long as the batch lasts.
- In *volume reduction,* operation is out of a finite reservoir into which the retentate stream is returned (batch); solute concentration of the feed thus increases systematically as permeate is withdrawn. Recovery is based on permeate volume relative to the original feed volume.

Intrinsic membrane performance (Table 3.2) is established under conditions of low recovery, operating at high feed flow rate and limited membrane area while restoring permeate and retentate streams into the feed reservoir.

Concentration polarization. – Less readily assessed in its effect on performance than permeate recovery, concentration polarization refers to the accumulation of rejected solute(s) near the feed-membrane interface, as pictured by the concentration profile shown in Fig. 4.3 (upper). Solute concentration build-up (*wall concentration over bulk concentration, c_w/c_b*) is represented by the thickness of the *laminar boundary layer* (δ), taken to be the distance over which a concentration gradient exists to effect back diffusion of the "trapped" solute(s) into the bulk feed stream. The adverse effects of concentration polarization are again inferred from the flux equations: Declining solvent flux and increasing solute passage on account of a higher than bulk concentration at the membrane surface.

As back diffusion out of the laminar boundary layer depends on the size of the diffusing species (Table 4.2), the effects of concentration polarization become more severe as the molecular size (molecular mass) of the rejected solute(s) increases. Since solubility at the same time decreases, the upper limit of concentration polarization is solute precipitation on the membrane surface (Fig. 4.3, lower). Salt precipitation when demineralizing "hard" feed waters (gypsum as a solute in point), referred to as *scaling*, may require to restrict permeate recovery. By *fouling* is meant the process of irreversible deposition of macromolecular matter on the membrane surface; *biofouling* refers especially to "natural organic matter", NOM (Table 3.3). – Fouling is a fact of life in *ultra- and microfiltration*, Sect. 4.2.

Concentration polarization in barrier separation cannot be avoided. When dealing with truly dissolved solutes, the only means to alleviate the problem is to influence the thickness of the laminar boundary layer through appropriate hydraulic process and apparatus (module) design. This is the origin of the *cross flow* ("tangential flow") mode of operation in barrier separation. When having to deal with particulate matter, a clarifying *pretreatment* of the feed stream is called for, which in itself may be a case of membrane filtration (Fig. 3.6). It is noted that waterborne pathogens tend to cling to suspended particulate matter.

3.2.4 Energy considerations

Pressure being the driving force in reverse osmosis, the energy expended is *electrical energy* to drive the high pressure feed pumps (recorded as kWh/m^3 of product water). The thermodynamic minimum energy required to separate pure water from a saline solution is readily established as the difference in free energy between pure water and the reduced free energy of water containing solute, that difference increasing with increasing solute (salt) concentration (Section 2.2.3). Everything beyond the minimum is "reality", to be assessed in terms of the three interrelated influence parameters –

- operating pressure $(p_0 \geq 2\pi)$;
- permeate recovery (Q_P / Q_0);
- concentration polarization $(c_w / c_b > 1)$.

Operating pressure (as inlet or feed pressure, p_0) is the only adjustable influence paramter and, as "pV work", is the principal energy consumer; regardless of recovery, *all* of the feed needs to be pressurized. *Specific energy consumption* refers to the fraction of the total which leads to product water as defined by the recovery of the process. The remaining energy is stored in the reject stream to be dissipated at the back pressure regulator (Fig. 3.5), which in effect is a throttle. Hydraulic *energy recovery* by means of special pumps or "pressure exchange" devices therefore is standard practice in high pressure reverse osmosis, and is in fact one of the few means to improve reverse osmosis economics.

Recovery has two implications in practical water desalination:

- When desalted permeate is withdrawn from the feed stream, the salinity of the remaining feed increases (volume reduction). Consequently, the *minimum free energy* of separation, even though defined to imply reversible (non-producing) conditions, *increases with recovery*. Clearly, this aspect is independent of the separation process used. Taking again "seawater" as example, the minimum free energy to produce pure water at room temperature is $0.7\,kWh/m^3$ (Eq. 2.14 with $x_2 = 0.018$ for "salt"). At a recovery of 50%, which is close to reality in seawater desalination, salinity of the reject stream about doubles; the minimum free energy of separation at this point would be $1\,kWh/m^3$.

- Under actual high pressure conditions and optimal recovery of product water, the energy of separation becomes specific energy of water production, outweighing the solute effect on the thermodynamic condition of the feed by far, and *decreases with recovery*. This aspect is peculiar to reverse osmosis in that practical recovery not only depends on feed water salinity, but can also be adjusted within limits: At low initial salinity, allowable recovery is high and required operating pressure low (and vice versa). This explains the advantage of reverse osmosis over distillation in demineralizing low salinity (brackish) feed waters.

Eq. 3.9 is an exercise in accomodating the dual effects of pressure and recovery on energy consumption in reverse osmosis. It is obtained by allocating pV work as equivalent electrical energy to the permeate fraction actually recovered, hence a specific energy consumption in kWh/m^3:

$$p_0 V = p_0 Q_0 h = p_0 Q_p h / E$$
$$\frac{kWh}{m^3} \approx 2.78 \, \frac{P[bar]}{E[\%]} \tag{3.9}$$

Complete recovery ($E = 100\%$) at osmotic pressure $p = \pi = 26\,bar$ formally registers as the minimum energy of separation for seawater, $0.7\,kWh/m^3$. Practical seawater reverse osmosis operates at pressures of 60+ bar and recovery of 40%, at a predicted energy consumption of 4+ kWh/m^3, – provided, that is, no hydraulic energy recovery is employed to reduce that figure. Again, lower feed salinity leads to lower energy consumption on both counts: Required pressure down, allowable recovery up.

3.2.5   Reverse osmosis in the real world

Reality is where the extra money goes. The undispensables of real-life reverse osmosis are sketched in Fig. 3.6, grouped into the three sections –

- feed water pretreatment;
- membrane section with pump and energy recovery;
- product water posttreatment.

Not considered are the feed water intake (well construction), nor local power supply, nor measures to dispose of the brine. Actual plant size, reported as daily product water capacity, ranges from less than 95 m^3/d = 25 000 gpd (innumerable units, not covered by international *Desalting Plants Inventory Reports*) to singular plants of capacity of the order of 50 000 m^3/d. There is, supposedly, little "economy of scale" to be realized, large plants being composed of lesser units in repetition, using common pre- and posttreatment.

Pretreatment. – Pretreatment has to meet with the adversities of local raw water conditions. The objective is to clear the feed stream from everything potentially harmful to membrane performance as evidenced by *flux decline* and/or undue limitation of *membrane life*. The measures taken are mechanical pre-filtration; acidification to reduce carbonate hardness, followed, if necessary, by aeration to reduce CO_2; addition of anti-scalants to keep divalent salts (sulfates) from precipitating; chlorination (or alternative oxidation) to deactivate microorganisms, usually followed by de-chlorination to safeguard against membrane degradation. In Fig. 3.6 two different pretreatment schemes are indicated, one of them using *membrane filtration* (MF or UF) in a design to eliminate pretreatment chemicals; membrane pretreatment would also result in substantial bacterial reduction.

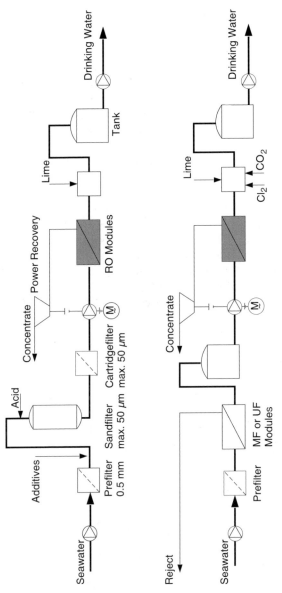

Fig. 3.6. The principal components of a seawater reverse osmosis plant, grouped in feedwater pretreatment; membrane plant including pump and energy recovery; product water posttreatment [6].

About half of the fouling deposit identified on "autopsy" of spent membrane elements, reported summarily in Table 3.3, is organic by nature (NOM), the other half being composed of low solubility inorganic species, – essentially everything mineral except monovalent salts. Distribution of foulants within a reverse osmosis stage (as schematized in Fig. 3.5) is uneven: While biofouling is most severe in the entrance region, danger of inorganic precipitation (scaling) increases with increasing recovery, that is, towards the end of the stage.

Posttreatment. – This is a lesson in water chemistry, Appendix B.

Devoid of all hardness components yet containing CO_2 from the acid pretreatment, demineralized water is *soft* and *acidic*, with the unpleasant and downright corrosive properties of that state. There is no principal difference in this regard between the *permeate* of reverse osmosis and the *distillate* from thermal desalination, except for the residual salinity of the permeate as "artifact" of membrane permeability, – mostly NaCl.

When the steam engine took to sea, seawater desalination became a necessity; the requirement was *boiler feed*, and that needed to be soft (to prevent scaling of heat exchangers). Today, desalination means *drinking water* and – increasingly so – *irrigation water* for specialized agriculture. In both uses the water has to meet certain regulations set forth, for example, by the *World Health Organization* (WHO). Posttreatment centers around *remineralization* with lime as it balances with CO_2 according to $Ca(OH)_2 + 2\,CO_2 \rightarrow Ca(HCO_3)_2$. Excess CO_2 beyond that needed to keep Ca^{++} (Mg^{++}) ions in solution

Table 3.3. Membrane fouling on a polyamide rejecting surface (spiral wound configuration): Foulants observed with 150 spent membrane elements from around the world. Data derived from [7].

Organics/biofouling	48.6	to	60.6	%
SiO_2	3.4		20.4	
Al_2O_3	1.4		6.3	
Fe_2O_3 ("oxide hydrate")	6.2		7.6	
$Ca_3(PO_4)_2$	1.6		13.4	
$CaCO_3$	1.5		4.8	
$CaSO_4$	1.4		3.4	
Unaccounted	11.1		15.1	

is *aggressive*, acting much like a mineral acid. Reverse osmosis membranes are permeable to dissolved CO_2; the permeate (like any soft water) tends to retain some excess CO_2 even after deacidification (by aeration). Lime treatment, therefore, not only conveys taste and compatibility with soap, but also controls aggressive CO_2 in protection of water distribution systems. – In case acidification is circumvented by relying on *membrane filtration* as pretreatment (Fig. 3.6, lower), CO_2 actually may have to be added along with lime to sustain the desired level of hardness.

3.3 The power of osmosis

3.3.1 The osmotic pump

Pressure increases the potential energy of anything, even if "anything" is incompressible for all practical purposes. Gravity increases the pressure of seawater with ocean depth at the well-known rate of about 1 bar for every 10 m of water column, nominally matching the osmotic pressure at a depth of 258 m (referring to "standard seawater": 3.45 w-% TDS; osmotic pressure 25.5 bar; 25°C). Salinity not only accounts for the osmotic pressure but also for an increase in density: Seawater is about 3% denser than fresh water at all pressures (testifying to the structuring power of ion hydration). Also seen as independent of pressure, that is, of gravity is the composition of seawater (= ratio of salts to water), implying constant osmotic pressure throughout. The model sketched is that of a *uniform ocean*, an off-shore approximation by which deep-sea fish seem to flourish.

The *osmotic pump* is a lesson in sportive science based on the above scenario [8]. A sturdy pipe, capped at one end by an ideally semipermeable membrane, is lowered membrane-first into the ocean. At a depth of some 258 m, fresh water starts to seep into the pipe by reverse osmosis. On further lowering, a fresh water column will rise inside the pipe. Now, if the osmotic pressure equilibrium across the membrane is to be maintained, the fresh water inside the pipe will have to rise faster than the pipe is gaining in depth, to accomodate the pressure gradient due to the difference in densities between fresh water (above the membrane) and saline water (underneath). Eventually, fresh water will reach the surface. The depth (length of pipe immersed) at which this happens may be estimated as follows (index 1 = fresh water; index 2 = seawater).

By fluid statics, the pressure gradient of a vertical liquid column (dp/dz) is a function of its density (ρ) and of given gravity $(g = $ acceleration of gravity),

$$dp = \rho_1 g \, dz \tag{3.10}$$

Let z be the height of the fresh water column at point of overflow, equal to the total vertical length of pipe immersed. At this point, the seawater pressure deep down balances the pressure of the fresh water column *plus* the osmotic pressure increment, visualized as virtual addition ($z' = 258$ m) to the fresh water column:

$$\rho_2 g z = \rho_1 g z + \rho_1 g z'$$

$$z = \frac{\rho_1}{\rho_2 - \rho_1} z' \tag{3.11}$$

Accepting, for simplicity's sake, seawater to be more dense than fresh water by a flat 3% throughout, osmotic pumping will commence at a depth of $z = 8600$ m.

3.3.2 Osmotic power generation

Osmotic power is hydroelectric power gained by utilizing *direct osmosis* to enlarge the volume flow of pressurized saline water delivered to a turbine. Operating at the junction of saline water (seawater) and a fresh water source, the hydraulics of the process are described as follows (Fig. 3.7). A saline feed stream (inlet flow rate Q_o) is pumped in cross flow fashion through a membrane installation at an hydraulic pressure of approximately half its osmotic pressure (meaning 10 to 15 bar in case of seawater), paralleled across the membrane by a fresh water stream at ambient pressure. Osmosis (osmotic influx) increases the "pV energy" of the saline feed stream by increasing its volume, diluting it in the process. The exit stream (exit flow rate $Q_e > Q_o$), still under pressure, is divided: One fraction is directed to the *turbine* to generate power; the second fraction, on its way to being returned to the sea, passes through a *pressure exchanger*, there to convey its pressure to the incoming seawater feed stream. The balance between liquid streams is such that it is the

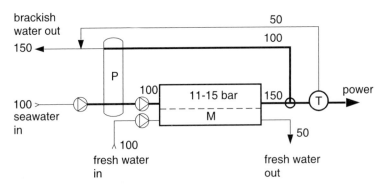

Fig. 3.7. Osmotic power generation, process scheme. P = pressure exchanger; M = membrane osmotic plant; T = turbine. Flow rates indicated are normalized to a saline water intake of $Q_o = 100$.

volume of water gained by osmosis which – at the preselected pressure – is available for power generation. The return flow thus is of equal magnitude, albeit lower salinity, as the seawater intake, reminiscent of a "feed-and-bleed" loop. The fresh water return, except for having lost half its volume, is as fresh as before.

Other than the osmotic pump, osmotic power generation may yet come true. A basis for feasibility studies is the preselected pressure head of 10 to 15 bar translating into a water column of 100 to 150 m which, when utilized in a hydroelectric plant, would amount to about 1 MWs of power per m² cross section of water column. Consequently, osmotic power is expressed in terms of W/m² of installed membrane area, thereby linking the effect to the flux performance of the membrane. At a goal of 5 W/m² – using seawater as the saline feed –, 200 000 m² of membrane area are slated for every MW of "installed" osmotic power.

Osmosis as a means for energy production was proposed by Loeb (1975), who used the term *pressure retarded osmosis* (PRO) to indicate that osmotic volume flux is designed to function against prepressurized saline water, that pressure constituting the operating pressure for the hydraulic power scheme [9]. Clearly, the net potential for osmotic power generation increases with the osmotic pressure of the saline feed solution. Small wonder, Loeb (who is also co-inventor of the asymmetric cellulose acetate membrane) had the Dead Sea in mind when proposing PRO.

Bibliography

[1] F. Rüdorff: Grundriss der Chemie, 15. Auflage. Verlag H. W. Müller, Berlin 1909.

[2] R. W. Field, loc. cit. Chap. 1, Ref. [11]. – S. Cross, Achieving 60 °Brix with membrane technology. 49th Annual Meeting, Institute of Food Technologists, New Orleans, 1988.

[3] H. K. Lonsdale, U. Merten, R. L. Riley, Transport properties of cellulose acetate osmotic membranes. J. Appl. Polymer Sci. 9 (1965) 1341–1362. – U. Merten (ed.): Desalination by Reverse Osmosis. MIT Press, Cambridge, MA, 1966.

[4] J. E. Cadotte, Evolution of composite reverse osmosis membranes. In: Materials Science of Synthetic Membranes (D. R. Lloyd, ed.), ACS Symposium Series No. 269, American Chemical Society, 1982.

[5] Data courtesy M. Wilf. See also M. Wilf, C. Bartels, Optimization of seawater RO systems design. Desalination 173 (2005) 1–12.

[6] International Atomic Energy Agency, IAEA–TECDOC–574, Vienna, 1990.

[7] M. Fazel, E. G. Darton, Performing a membrane autopsy. Desalination & Water Reuse 11 (2002) 40–46.

[8] O. Levenspiel, N. de Nevers, The osmotic pump. Science 183 (1974) 157–160.

[9] S. Loeb, Osmotic power plants. Science 189 (1975) 654–655.

4 Membrane Filtration

There is good reason to believe that filtering (= straining) is as old as brewing, Fig. 4.1. The medieval verb *filtrare* relates to *feltrum*, meaning anything compacted to serve as filter medium; *felt* is compacted wool or hair ("nonwovens" in filtration parlance).

4.1 On size and size exclusion

Filtration is convective discriminating mass transport of liquid mixtures or gaseous dispersions (aerosols) through porous barriers, mass transport ideally being confined to the void space of the barriers. *Sieving* refers to filtration of particulate matter; *gaseous diffusion* is the term used when all components are gases. Discrimination is by size. The common permeant in liquid membrane filtration is the solvent: Water.

In aqueous membrane filtration *effective* (observed) solute size usually differs from *geometric* (predicted) size as a result of interactions between the solutes (which may be charged) and the barrier

Fig. 4.1. Time-tested: Gravity assisted dead end filtration in ancient Egypt.

surface contacted (which may include the pore walls). Rarely there is a snug fit between solute and pore on purely geometrical terms.

In terms of molecular mass, solute size encountered in membrane filtration extends over five orders of magnitude, Table 2.1 and Appendix C present examples. Whether truly dissolved or microscopically dispersed (oftentimes a matter of semantics), pore sizes to restrain solute passage are such that the force of gravity (Fig. 4.1) no longer suffices to overcome the *hydraulic resistance* of the barrier. Membrane filtration, accordingly, is pressure driven barrier separation of aqueous solutions, loosely grouped into a number of process variants with reference to the size brackets of the solutes handled:

- nanofiltration (NF)   0.01–0.001 µm   (< 10 nm);
- ultrafiltration (UF)   0.2–0.005 µm   (5–200 nm);
- microfiltration (MF)   10–0.1 µm   (> 100 nm).

Classifying solutes follows these categories. By a geometrical argument based on solute diffusivity in aqueous solution (Table 4.2), solute diameter of nonelectrolytes relates to solute molecular mass approximately thus:

$$d[nm] \approx 0.13 \times MW^{1/3} \tag{4.1}$$

As a token of reference, a solute diameter of 1 nm is roughly equivalent to a molecular weight of 500 g/mol; *microsolutes* below this order register as "osmotically relevant". *Macromolecules*, represented in the main by proteins, polysaccharides, natural rubber, synthetic polymers, are characterized by their molecular mass. The spectrum of *marker molecules* presented in Appendix C describes the mass range of applied ultrafiltration. *Colloids* are macromolecules in aqueous dispersion, and are described by their *effective size*; in fact, most anything dispersed in water before becoming optically detectable, that is, up to 200 nm (0.2 µm), is termed colloidal. If freely mobile (dilute), dispersed macromolecules are referred to as *sol*; when becoming "entangled" (concentrated), they turn into a *gel*. Still bigger by another order of magnitude, and definitely particulate, are microorganisms and biological cells, living or dead (cell debris), assigned to the 1 to 20 µm size range.

It is mostly *proteins* which are to be retained or fractionated by ultrafiltration. Proteins are charged biopolymers; when viewed as aqueous solutes, they are pictured as composed of a hydrophobic "core" into which the sequence of constituent amino acids is folded, outwardly studded with amino groups (positively charged) and carboxylic acid groups (negatively charged). Charge interaction with the membrane tends to increase the effective size, resulting in *electrostatic retention* to be higher than *size retention*.

Figure 4.2 depicts the operative range of pressure driven membrane processes from reverse osmosis to ordinary particle filtration: Operating pressure (indicative of hydraulic barrier resistance) relative to solute size to be retained. *Reverse osmosis*, phenomenologically a filtration effect ("hyperfiltration"), more appropriately is interpreted as a solution-diffusion process on account of the prevailing mode of barrier interference (Chapter 3.2).

What may appear like an orderly development towards ever more delicate filtration capability in fact conceals a very different genesis. Up to the time of Ferry's 1936 review on *ultrafiltration* (Chapter 2, Ref. [5]) the only "synthetic" film forming polymers were cellulose

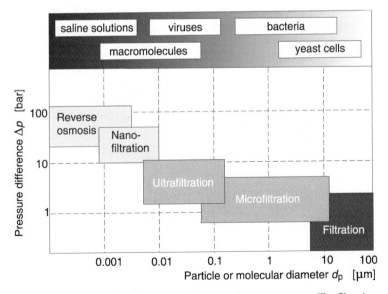

Fig. 4.2. Operative range of pressure driven membrane processes: The filtration spectrum. Increasing operating pressure signifies increasing hydraulic resistance of the barriers employed. From [1] with permission.

esters (*collodion*, Sect. 6.1) and regenerated cellulose (*cellophane*), artfully prepared into membranes of graded porosity for laboratory use. Pore sizes from 5 μm down to 5 nm are reported, filtration of pathogens was of foremost interest, flux was of no concern. Water flux came into focus when, in the 1950's, Reid, using dense cellulosic films as thin as laboratory art could make them (< 10 μm), demonstrated that strong electrolytes (salts) could be pressure-filtered, the filtrate being demineralized water. Water flux in what was to become *reverse osmosis*, however, remained too low to be of economic interest, Reid gave up.

Two events fostered the subsequent development:

- The discovery of the asymmetrically structured and thereby effectively thin cellulose acetate membrane (Loeb and Sourirajan, 1960; Sect. 2.3.3), which opened the way to practicable water fluxes under reverse osmosis conditions.
- The surge of polymer technology after World War II, accompanied by a surge in polymer film technology. In particular, the structural principle of the asymmetric membrane soon was applied to other polymers as well, creating synthetic microporous membranes with a favorable aspect ratio (pore length to pore diameter) in the process. What *cellulose acetate* was for reverse osmosis came to be *polysulfone* for ultrafiltration (Chapter 6).

Practical ultrafiltration (since 1965) thus is a follow-up to practical reverse osmosis, as is practical gas separation (since 1980). Nanofiltration, originally an ill-appreciated foundling, now bridges the gap between ultrafiltration and high retention reverse osmosis (a size bracket into which virus and multivalent ions belong).

4.2   Liquid transport in membrane filtration

The model image of a porous barrier is a perforated sheet with straight cylindrical pores of uniform diameter extending vertically to the plane of the sheet: an "isoporous sieve". The *surface porosity* (open area) of such a barrier is $\pi r^2 n$, its *volume porosity* (void volume) is $\pi r^2 n z$ (n = number of like pores; z = length of pore, equal to barrier thickness). It is noted that, in terms of void fraction, surface porosity conforms to volume porosity.

Nowhere is reality so much more interesting than the model as in the case of microporous structures. A pictorial record of the structural diversity of real porous barriers – polymeric, mineral, metallic – is presented in Appendix E. As to polymers, it is noted that only "glassy" or else highly crystalline polymers provide the structural integrity required to maintain pores in liquid contact (Sect. 6.2). The message at this point is that fairly well-defined molecular solutes are being matched with the *pore size distribution* of actual barriers.

Nevertheless, assessing mass transport in and about porous barriers retains the descriptive simplicity of the model barrier, adding hindsight refinement to accommodate reality as needed. Central to interpretation is the convective liquid volume flux (J_v) as it is influenced by the presence of solutes. The following situations need to be considered:

- volume flux of pure water;
- sub-critical flux from aqueous solutions (partial solute retention);
- critical flux from aqueous solutions (complete solute retention); and
- transport behavior of permeable solutes.

Pure water flux. – As *hydraulic permeability* pure water flux is one of the parameters characterizing porous barriers. The range is considerable; water flux commensurate with the porosities covered by the *filtration spectrum* ranges from 10 to > 1000 L/h m^2 (corresponding to convection velocities of between 1 and > 100 cm/h). For perspective, water flux at the 10 gfd threshold of reverse osmosis is 17 L/h m^2 at prevailing pressure (Table 4.1).

An expression for water flux through pores is derived from the Hagen-Poisseuille equation describing hydraulic pressure loss within a capillary duct (laminar flow) as

$$J_v = \frac{\varepsilon d^2}{32 \eta \tau} \frac{\Delta p}{z} = L_p \frac{\Delta p}{z}$$

with geometrical porosity   (4.2)

$$\varepsilon = \frac{n \pi d^2}{4}$$

the trivial message being that there is no convective flux at $\Delta p = 0$ (and that the flux diminishes with increasing length of the duct,

Table 4.1. Performance of ultrafiltration membranes: Water flux and rejection of marker molecules as function of pore size rating. Reproduced from [2].

Nominal MW cutoff	Apparent pore diameter (nm)	Water flux at 3.7 bar (L/h m^2)	Rejection (%)			
			D-Alanin	Sucrose	Myoglobin	IgM
500	2.1	17	15	70	>95	>98
1000	2.4	34	0	50	>95	>98
10000	3.0	102	0	25	95	>98
10000	3.8	935	0	0	80	>98
30000	4.7	850	–	0	35	>98
50000	6.6	425	–	–	20	>98
100000	11	1105	–	–	–	>98
300000	48	2215	–	–	–	>98

D-Alanin, MW 89
Sucrose, MW 342
Myoglobin, MW 17500
IgM (Immunoglobulin), MW > 900000

which is the original teaching of Hagen-Poisseuille). Parameters determining hydraulic permeability L_p according to Eq. 4.1 are: The surface porosity of the model membrane ($\varepsilon = n\,\pi d^2/4$); the viscosity of the liquid feed *within* the pore space, η; an adjustable parameter τ which symbolizes the fact that real pores are neither straight nor uniform ("tortuosity factor"). Flux is seen to depend on pore diameter to the fourth power, greatly amplifying the influence of the pore size distribution towards higher than mean pore rating.

4.2.1   Concentration polarization

The scenario, unfolded in Table 4.1, is one of increasing pore size rating (a vocabulary such as "nominal" and "apparent" indicating leeway) meeting specified solutes in order of increasing molecular mass. The aim of the following discussion is to describe liquid volume flux in membrane filtration (J_v) by considering the transport behavior of the solutes (J_i), which thereby assume the role of marker species: The *polarization model* (also known as *stagnant film model*). Refer to Fig. 4.3 for illustration and symbols used.

The smallest solutes encountered are ordinary salts, – and urea (Table 2.1). They are presumed to pass freely in ultrafiltration

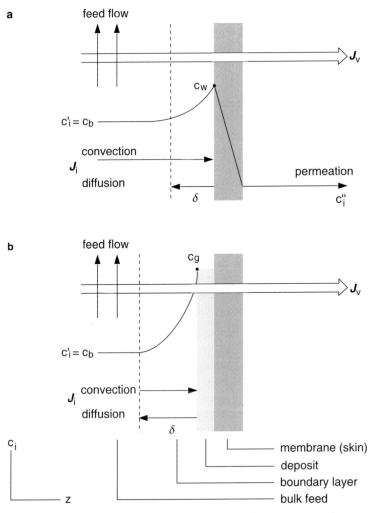

Fig. 4.3. The polarization model (film model) in cross flow membrane filtration. Upper: concentration polarization; lower: gel polarization.

$(c_i'' = c_i')$, while being somewhat retained by nanofiltration membranes $(c_i'' < c_i')$. Practical implications hinted at are: The use of microporous membranes to demineralize macromolecular solutions, – and to treat uremia by hemodialysis (Sect. 4.5).

As soon as some fraction of the solute is denied passage (by whatever mechanism invoked), several interrelated things happen (Fig. 4.3, upper).

- Solute concentration in the vicinity of the rejecting surface increases above feed level: *Concentration polarization*. As a result an osmotic pressure gradient develops across the membrane, reducing the available pressure head for liquid flux. The effect is basically similar to what is observed in reverse osmosis (Sect. 3.2.3), and is represented by an analogous flux equation as

$$J_v = L_p (\Delta p - \sigma \Delta \pi) \tag{4.3}$$

(compare with Eq. 3.4 of the Merten model). Equation 4.3 describes partial solute rejection by adjusting the influence of osmotic pressure through a *reflection coefficient* ($\sigma = 0$ to 1), which is patterned after the salt rejection of reverse osmosis. Even though osmotic pressures created by macromolecules are small, when accumulating they may reach the same order as the pressure applied in ultrafiltration.

- As a further result, the concentration gradient developing between wall concentration (c_w) and feed (c_i') incites a process of back diffusion counter to the direction of convective solute flow. The realm of back diffusion into the turbulent feed stream defines the *laminar boundary layer* δ, its thickness taken as indicative of the degree of concentration polarization, $c_w/c_b > 1$. The term *bulk concentration* c_b is introduced to identify the turbulent feed flow regime under conditions of concentration polarization; it corresponds to the concentration of the "well mixed" feed stream.

Steady state is characterized by a balance of solute flow complements, in which the total convective solute transport towards the membrane ($= J_v c_b$) is balanced by the fraction permeating ($= J_v c_i''$) plus the fraction returning to the bulk stream by (Fickian) diffusion ($= -D_i\, dc_w/dz$). On integration the *polarization equation* is obtained, derived in the early days of reverse osmosis [3],

$$\frac{c_w - c_i''}{c_b - c_i''} = exp\left(\frac{J_v \delta}{D}\right) \tag{4.4}$$

Table 4.2. Indicative diffusion coefficients of nonelectrolytes in aqueous solution as function of molecular size. Derived from [4].

Molecular weight (g/mol)	Diameter (nm)	Diffusion coefficient (cm^2/s)
10	0.29	2.20×10^{-5}
100	0.62	0.70
1000	1.32	0.25
10000	2.85	0.11
100000	6.2	0.05
1000000	13	0.025

d [nm] $\approx 0.13 \times MW^{1/3}$

The dimensionless composite $J_v \, \delta/D$ represents the ratio of convective to diffusive solute transport in the boundary layer. In a nutshell it contains all the information needed to describe the dynamics of the membrane-solute system under consideration:

- The volume flux J_v relates to membrane porosity, high porosity leading to high convective flux, in turn causing the wall concentration c_w to increase, in turn causing the thickness of the boundary layer δ to grow.
- The diffusion coefficient D indicates solute molecular size, which (a) determines the diffusive mobility, and (b) limits the solubility of the solute species in water. Table 4.2 lists indicative diffusion coefficients of nonelectrolytes over the size range encountered. It is noted that the molecular diameter merely doubles as the molecular mass increases by an order of magnitude.

However: what is the solubility limit of colloidal macromolecules?

4.2.2 Gel polarization

There is no ordinary limit to the concentration of macromolecules in aqueous solution (as would entail a separation into precipitate and supernatant), however, there is a *limiting situation*: *Gelation*. Gelation is dewatering of the macromolecular solution to the point where it solidifies. In ultrafiltration that point is reached next to the rejecting surface (the "wall") in the limit of complete solute rejection, producing a gelatinous deposit which constitutes a secondary barrier to liquid volume flux on top of the original membrane (Fig. 4.3, lower). In practical ultrafiltration of macromolecular solutions (imagine cheese whey), gel formation occurs within minutes, progressing from entrance to exit region of the membrane stage, and is largely irreversible.

In the formalism of the polarization model (Eq. 4.4) gelation shows up as zero permeate concentration of the gel forming solute ($c_i'' = 0$), and is signified by replacing the wall concentration by a hypothetical gel concentration ($c_w \rightarrow c_g$), – gel polarization [5]:

and
$$\frac{c_g}{c_b} = exp\left(\frac{J_v \delta}{D}\right)$$
$$J_v = \frac{D}{\delta} \ln\left(\frac{c_g}{c_b}\right)$$
(4.5)

Assessing gel polarization again is guided by the volume flux J_v, observing that – as a rule – the deposit is less permeable to water than the porous barrier itself (the notion of a "dynamic membrane"). The key quantity is the *critical flux*, defined as the lowest flux at given solute concentration to result in an irreversible deposit on the barrier surface. When followed as function of gradually increasing pressure, flux increases until, at the occurence of complete gel layer coverage, becoming independent of pressure (Fig. 4.4). If pressurized further, the flux remains at the critical level, either by the deposit growing thicker or by mechanical compaction; compaction of the gelatinous deposit eventually may even cause the flux to decline with pressure.

Equation 4.5 summarizes the parameters which determine critical flux (respectively the conditions to avoid gel layer coverage): Solution-related (c_b and D); process-related (δ). In turn:

- Volume flux is predicted to decrease with the logarithm of bulk feed concentration, c_b. That dependence reflects the effect of increasing permeate *recovery* along the length of a membrane stage (Fig. 3.5), or else in batch dewatering, – up to the fictitious limit of bulk gelation.

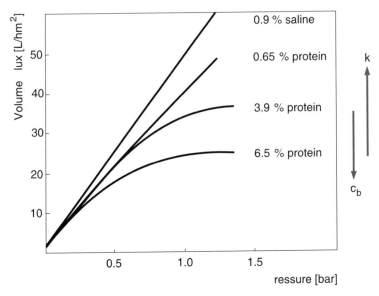

Fig. 4.4. Ultrafiltration: Effect of pressure, solute bulk concentration (c_b) and hydrodynamic condition (k) on flux. Saline: isotonic salt solution (Sect. 3.1.3); protein: albumin. Adapted from [5].

- The dependence of flux on solute diffusivity is a consequence both of solute *solubility* affecting its inclination to gel and of solute *mobility* affecting back diffusion across the dynamic boundary layer.
- The ratio of diffusion coefficient and boundary layer thickness, $k = D/\delta$ (a "velocity" by dimension), is a *mass transfer coefficient* describing the combined influence of solute dynamics (diffusion coefficient) and fluid dynamics (cross flow rate) on solute transport. Mass transfer coefficients are obtained summarily as slopes of the flux versus log-bulk concentration curves, without specific reference to solution properties (Eq. 4.5).

Microfiltration is different. – As solute size (and incumbent pore size) increases beyond the realm of bona fide solubility, the scenery changes. Within a dimensional reach of approximately 100 to 1000 nm (0.1 to 1 μm) a transition occurs from colloidal solution to particulate dispersion. Indeed, it is not only biological material, witnessing the transition from macromolecules to microorganisms, but also waterborne inorganic species like clay components, silica, oxide-hydrates of heavy metals, and hitchhiking humic acids, which make their appearance as filterable solutes. The *silt density index* (SDI), which routinely records reverse osmosis feed water turbidity, is based on microfiltration through a 0.45 μm standard membrane filter.

As the nature of the solute changes, so does the conformation of the membrane deposit. What is observed is a *higher* permeate (filtrate) flux than predicted by simple gel polarization as solute size increases and solute response to hydrodynamic conditions changes. In terms of the parameters of the gel polarization model (Eq. 4.5) it appears that mass transfer coefficients increase abnormally, considering that aqueous diffusivities are supposed to even decrease with molecular mass (Table 4.2). Several mechanisms to explain the enhanced transport of high molecular weight solutes away from the membrane surface are being considered, among them *hydrodynamic lift forces* (primarily affecting D) and *shear-induced diffusion* (primarily affecting δ), Eq. 4.5.

4.3   Solute transport in membrane filtration

The issue is *partial solute rejection* in ultrafiltration (as exemplified in Table 4.1); it is a subject matter of considerable research attention and never-ending speculation. Key terms in the discussion are *pore blockage* and *pore constriction* [6]. When a singular solute species is partially rejected, irregularities in membrane pore structure (pore size distribution) and site-specific solute-polymer interactions are held responsible. If solutes differing in size are to be fractionated, barrier irregularities and molecular interactions add to size discrimination, limiting the fractionating prowess of ultrafiltration. As a rule, to effect membrane protein fractionation requires molecular sizes to differ by nearly an order of magnitude.

The gross solute separation capability of membrane filtration is addressed thus: *Ultrafiltration* retains macromolecules while being freely permeable to microsolutes. *Microfiltration*, in turn, permeates macromolecules while retaining microorganisms and cells. Specific applications follow the gross pattern.

Dialysis. – Dialysis is diffusive mass transport transmitted by the pore fluid, and thus pertains to the fraction of solute which actually enters the pore space of the membrane. The driving potential is linked to the solute concentration gradient between "entrance" and "exit" of the pores, the exit concentration being identical to that of the receiving phase (now called *dialysate*). Undesired osmotic flux of solvent (in either direction) is quenched by maintaining *isotonic conditions* on both sides of the membrane.

If the dialysate is continually renewed, the feed solution gradually becomes depleted of the permeating solute, – it is "washed out". Not to exhaustion, to be sure, as the diffusion gradient diminishes along with depletion; dialysis, as any concentration-driven mass transfer, slows down asymptotically.

The obvious application of *depletion dialysis* is to free macromolecular solutions of unwanted microsolutes under gentle operating conditions; a conspicuous application is *hemodialysis*.

Diafiltration. – As witnessed by the duration of hemodialysis treatment, dialysis is slow, and confined to low molecular weight solutes. When a pressure is imparted on the feed solution, a convective flow is added which is considerably more effective in transporting solutes than diffusion alone. The process, now a hybrid of dialysis and ultrafiltration (Eq. 4.6), concentrates the feed solution through *volume reduction*, which, unless wanted, may require judiciously replenishing the solvent lost into the dialysate.

$$J_i = J_v \overline{c}_i - D_i \frac{d\overline{c}_i}{dz} \tag{4.6}$$

(the bar to denote an average solute concentration within the pore fluid). As the feed pressure is raised, pore flow turns into straight convection and the concentration gradient within the membrane disappears: *Plug flow*. Solute separation in diafiltration under plug flow conditions is equivalent to sieving at the porous membrane surface. A necessary corollary in case of *partial solute rejection* is again concentration polarization.

4.4   Rating porous membranes

By a rule of error analysis, uncertainties multiply. When microporous barrieres are to be characterized (rated), the uncertainties in question are the morphology of the barrier itself and the size assay of the macrosolutes with which the membranes are challenged. If rarely there is a rational fit between pore size and solute size, it is the influence of solute-polymer interaction in addition (it is recalled that proteins are charged). Nevertheless, it is the *observed* ability to retain macromolecules (by ultrafiltration) or microorganisms (by microfiltration) which characterizes porous membranes.

Ultrafiltration membranes are described by *nominal ratings*. The molecular weight cutoff (MWCO) is defined as the molecular weight of a test solute (preferably a globular protein) which is 90% retained by the membrane in question. The molecular weight range of "marker molecules" (Appendix C) conforms to the range of applied ultrafiltration, which extends from glucose (MW 180; permeable) to immunoglobulins (MW > 900000; impermeable). To establish the rejection profile of an ultrafiltration membrane, a number of marker molecules is selected to cover the entire rejection range of that membrane; from the resulting curves of rejection versus molecular weight the MWCO at 90% rejection is obtained graphically (and absolute rejection is inferred), Fig. 4.5. As would be expected, the sharpness of the rejection profile is a reflection on the narrowness of the pore size distribution; further, at comparable mean rejection, the nominal (and absolute) MWCO assumes higher values as the rejection profile becomes more diffuse.

Sieving and clearance. Filtration seeks to separate solute from solvent, ultrafiltration moreover solute i from solute j in the process. If the solute is unwanted (as *salts* in reverse osmosis), the term is "rejection"; if the solute is wanted (as *proteins* in ultrafiltration), the term is "retention". Whereas the MWCO characterizes individual membranes, the *sieving coefficient* addresses individual solutes as ratio of solute concentration in permeate (filtrate) and feed: $S_i = c_i'' / c_i' = 1 - R$ (compare with Eq. 3.7). When applying ultrafiltration to discriminate between differently sized solutes, a *separation factor* is formed: $\alpha = S_i / S_j$ (analogous to Eq. 5.5). If the smaller solute is freely permeable (marked to be removed), $S_i = 1$, and the separation factor reduces to the reciprocal of the sieving coefficient

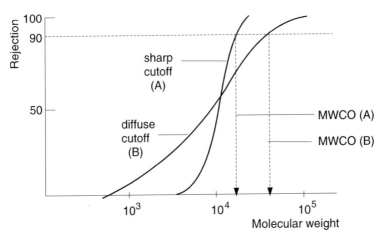

Fig. 4.5. Rejection profiles of two ultrafiltration membranes with indication of their nominal rating in terms of the 90% molecular weight cutoff. After [7].

of the retained macromolecule: $\alpha = 1/S_j$. In typical ultrafiltration practice, solutions (dispersions) of proteins and cells are to be freed of microsolutes.

A graphical *proof of consistency* of ultrafiltration performance is obtained by plotting the selectivity (as $1/S_j$) of a given marker molecule against the hydraulic permeability (L_p) of the membranes under consideration. When falling at or near the "trade-off" curve, membrane performance is considered adequate [8].

Hemodialysis (depletion dialysis) is highly selective solute-solute separation at constant feed volume (Sect. 4.5). In addition to selectivity, the mass transfer capacity of the membrane or hemodialyzer is needed. It is expressed as *clearance (plasma clearance)* in terms of the volume of blood which is completely cleared of a given uremic toxin per unit time (*mL/min*). Not only is clearance specific to each toxic species (to be established using appropriate marker molecules, Table 2.1), clearance values also depend on the mode of operation: Purely dialytic (diffusion) to purely plug flow (convection), Sect. 4.3.

Microfiltration membranes are described by *absolute ratings*. This rating is based on the notion of complete rejection of microorganisms in dead-end filtration, and thus specifically observes the largest pore. Presentation is in terms of graded pore diameters relative to the sizes of selected "marker microorganisms" (or substitute marker species) which are (completely) retained; standard denotations of microfiltration membranes based on marker filtration are 0.1 / 0.2 / 0.45 / 0.8 / 1.2 µm of (largest) pore diameter.

Microbial filtration, followed by cultivation to facilitate identifying and counting of the microorganisms rounded up, has become standard practice in drinking water sanitary assay. The method was developed in Germany during World War II, when city water supplies, endangered by bombing, needed rapid safety assessment.

4.5   Notable applications

Most membrane filtration operations are busily and profitably at work, and present no more challenge than available membranes (Appendix E) and supporting technology are able to cope with. Established applications are: Sterile filtration in medical and beverage operations; recovery of electrocoat paint; processing of milk products and fruit juices; process-specific wastewater treatment.

As to prevailing technique, ultrafiltration (like reverse osmosis) is operated in the cross flow mode (high ratio of feed flow rate over permeation rate). In microfiltration, the solids content of the feed determines process configuration: Cross flow at solids content > 0.5%, dead-end flow (the literal filtration mode) at solids content below that rule of thumb figure (Eykamp). Accordingly, sterile filtration operates as dead-end filtration.

Two ramifications of applied membrane filtration deserve attention beyond: The *membrane bioreactor* (MBR), its future role in bio-organic synthesis and wastewater management as yet unfathomed; the *artificial kidney* (hemodialysis, HD), its phenomenal achievement encouraging continued research on membranes in life-supporting systems.

Membrane bioreactor. – The essence of a membrane bioreactor is to conduct bioreactions (as synthesis or degradation) in direct proximity to a semipermeable barrier, technically combining the steps of reaction and product recovery with a hint at continuous operation. Although not a priori confined to any particular type of barrier, current membrane bioreactor development focuses on microporous membranes in their capacity to contain microorganisms. The two directions of *membrane biotechnology* feasible are bioorganic synthesis and biological wastewater treatment, respectively, – controlled *fermentation* of biomass both.

- *Bioorganic synthesis.* It is a matter of anticipation whether membrane biotechnology eventually will compete with petrochemistry in providing organic base chemicals. The aim is to improve on bioreactor productivity through removal of biotoxic metabolites, facilitating product recovery ("downstream processing") at the same time (Sect. 5.6).
- *Wastewater treatment.* Microbial decomposition of organic sewage, replacing the conventional activated sludge/settling treatment by an activated sludge bioreactor in combination with membrane filtration [9]. In addition to high quality effluent (water reuse), objectives of the MBR technology in progress are to reduce treatment time (reactor efficiency), quantity of sludge to be disposed of, and plant acreage ("foot print"). As would be expected, the single most pressing problem is membrane fouling [10].

Hemodialysis. – The human kidney processes about 1000 L of aqueous solution every week. Even though falling behind in exchange capacity, artificial membrane devices have come close to mimicking the clearance function of the kidney, which is to remove biotoxic metabolites from the blood stream. As a membrane separation process, hemodialysis is governed by the molecular mass (size) of the uremic toxins to be eliminated relative to the mass of serum proteins to be retained. Diffusion alone transports solutes up to MW 1000 (*hemodialysis* proper), and is slow. On applying pressure, convective transport is added, extending the mass transfer capacity to higher MW solutes (like β2-microglobulin, MW 11800), ultimately turning dialysis into ultrafiltration (*hemofiltration*) characterized by substantial elimination of solutes to MW 40000 while retaining essential proteins (serum albumin, MW 69000). Required solute fractionation thus is between MW 10000 and lower (permeating) and MW 60000 (retained), operating in an environment liable to protein fouling.

The classical membrane material for hemodialysis is *regenerated cellulose*, almost as nature provides it (Sect. 6.2). More recent developments use synthetic high T_g polymers (PES, PSU; Appendix D), which are hydrophilized by blending with polyvinylpyrrolidone (PVP). It appears that a barrier surface simultaneously exhibiting hydrophilic and hydrophobic functions is a viable answer to the problems of biocompatibility and protein fouling in general [11]. As in reverse osmosis and ultrafiltration, synthetic hemofiltration membranes are asymmetrically structured for flux (Fig. E.8); pore size distribution is narrow, centering around a pore diameter below 10 nm, which is in the nanofiltration range.

A typical hollow fiber dialyzer has a membrane area of $1.5 \, m^2$; a typical patient sufferung from chronic kidney failure requires 150 treatments per year. At an estimated one million individuals so afflicted the membrane area to come into contact with life blood amounts to well over 200 million m^2 annually, at a price. It deserves mention that the proceeds of this beneficial endeavor have inspired a scientific award (the *Crafoord Prize*) to promote basic research in areas which the *Nobel Price* does not recognize: Mathematics, geosciences, biosciences, astronomy.

Bibliography

[1] T. Melin, R. Rautenbach, loc. cit. Chap.1, Ref. [10].

[2] Source: Amicon Corporation. Reproduced by M. Cheryan, M. A. Mehaia in: Membrane Separations in Biotechnology (W. C. McGregor, ed.), Marcel Dekker, New York, 1986.

[3] P. L. T. Brian in U. Merten (ed.): Desalination by Reverse Osmosis. MIT Press, Cambridge, MA, 1966.

[4] R. H. Perry, C. H. Chilton: Chemical Engineer's Handbook, 5th Edition. McGraw-Hill, Auckland, 1974.

[5] W. F. Blatt, A. Dravid, A. S. Michaels, L. Nelson, Solute polarization and cake formation in membrane ultrafiltration. In: Membrane Science and Technology (J. E. Flynn, ed.), Plenum Press, New York, 1970.

[6] For example: C. C. Ho, A. L. Zydney, A combined pore blockage and cake filtration model for protein fouling during microfiltration. J. Colloid Interface Sci. 232 (2000) 389.

[7] M. Mulder, loc. cit. Chap. 1, Ref. [8].

[8] A. Mehta, A. L. Zydney, Permeability and selectivity analysis for ultrafiltration membranes. J. Membrane Sci. 249 (2005) 245–249.

[9] S. Judd: The MBR Book, Principles and Applications of Membrane Bioreactors in Water and Wastewater Treatment. Elsevier, Oxford, 2006.

[10] P. Le-Clech, V. Chen, A. G. Fane, Fouling in membrane bioreactors used in wastewater treatment. J. Membrane Sci. 284 (2006) 17–53.

[11] M. Storr, R. Deppisch, R. Buck, H. Goehl, The evolution of membranes for hemodialysis. In: Biomedical Science and Technology (Hincal, Kas, eds.), Plenum Press, New York, 1998.

5 Pervaporation versus Evaporation

5.1 Phenomenon and realization

Quote: *In the course of some experiments on dialyzation, my assistant, Mr. C. W. Eberlein, called my attention to the fact that a liquid in a collodion bag, which was suspended in the air, evaporated, although the bag was tightly closed.* This is the original observation reported by Kober in 1917 (Chap. 7, Ref. [12]). – Collodion is cellulose nitrate, permeable to water (Sect. 6.1). A parallel to the loss of water through packaging film or, for that matter, contact lenses comes to mind.

Pervaporation is mass transfer from liquid to vapor across interactive permeable barriers. When applied to volatile liquid mixtures, pervaporation results in a separation effect to be likened to that of distillation. However, whereas selectivity in distillation is predictable by the rules of equilibrium evaporation alone (vapor-liquid equilibrium, VLE), pervaporation additionally is influenced by specific membrane-solvent interaction (barrier interference) which provides access to unconventional ("difficult") liquid separation effects.

The diagram of an experimental set-up in Fig. 5.1 illustrates the working principle. The thermostated liquid feed is pumped at ambient pressure (p') across a membrane module, the reject stream being recirculated into the reservoir. Pervaporation is effected by maintaining a reduced pressure at the downstream side of the membrane by means of a combination of cold trap and vacuum pump, causing the permeate to evaporate as it emerges from the membrane (at downstream pressure p''). The vaporized permeate is recondensed in the cold trap.

In Kober's experiment, a reduction of partial pressure of the vaporizing permeate is achieved through dilution with air; "carrier gas pervaporation" would be the corresponding process realization.

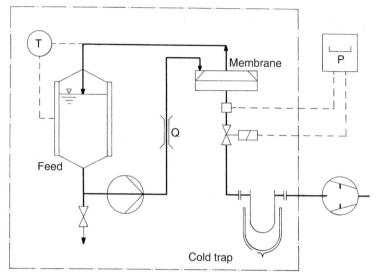

Fig. 5.1. Experimental set-up of vacuum pervaporation. T = temperature; Q = feed flow rate; P = downstream pressure. Recirculating the reject stream systematically extracts the "faster" feed component: Batch operation (akin to depletion dialysis).

5.2   Mass transport and selectivity

Candidate liquid mixtures are aqueous-organic solutions as characterized in Sect. 2.1.2, the variant component being the multitude of organic "solvents" (VOC's = volatile organic compounds) coming as *high boilers* or *low boilers* relative to the normal boiling temperature of water. For transport analysis, the solution-diffusion model (Sect. 2.3) is invoked, presupposing homogeneous polymer membranes. This model is not wholly representative, as follows from the fact that water, the smallest of liquid molecules at room temperature, effectively pervaporates through inorganic microporous barriers (zeolites; Fig. 5.3). Also, Kober's collodion bag usually counts as mildly swollen microporous (cellulose nitrate, Sect. 6.1).

Single component mass transport in terms of the solution-diffusion model is represented by the following equations, repeated from Chapter 2:

Flux
$$J_i = \frac{c_i^m D_i}{RT\,z}\,\Delta\mu_i$$
(5.1/2.16)

Driving force
$$\Delta\mu_i = \overline{V}_i\,\Delta p + RT\,ln\frac{p_i'}{p_i''}$$
(5.2/2.13)

It is noted that *single component flux* may refer to a pure liquid permeating (p_i° = saturation vapor pressure) or to the partial flux of a specified component of a liquid mixture (p_i' = partial pressure in feed mixture), – turning into vapor when emerging from the membrane. Adaption of the solution-diffusion scheme to the peculiar circumstances of pervaporation is as follows.

Permeability. – The membrane finds itself exposed to liquid feed and gaseous permeate. Accordingly, sorption and swelling is fully developed at the feed side only, whereas the permeate side is essentially "dry". Sorption and the sorption profile across the membrane depend on polymer-solvent interaction. Under steady state operating conditions, the parameters determining membrane permeability – permeant solubility (c_i^m), permeant diffusive mobility (D_i), and membrane thickness (z) – are no longer predictable without additional information on the sorption situation. Two limiting situations, represented by two classes of permeable barriers, can be identified:

- *Low total sorption* (linear or *Henry* sorption isotherm). The polymers are highly crosslinked ("stiff"), swelling and permeability are correspondingly low with a definite preference for small permeant molecules, – water in particular. The polymers are referred to as *glassy* or *semicrystalline*, their ideal realization being inorganic molecular sieves (zeolites), which do not swell at all. – The hydrophilic CA and PA membranes employed in reverse osmosis (Table 3.2) are of the glassy type.
- *High total sorption* (positively nonlinear or *Flory-Huggins* sorption isotherm). A low degree of cross-linkage facilitates membrane swelling extending far into the membrane, in turn enhancing permeant mobility on the whole (Sect. 2.3). The polymers are

referred to as *rubbery* or *elastomeric*, preferential interaction is with organic solutes. The ideal realization are liquid membranes (in the form of supported liquid membranes, SLM) which, by virtue of their being water insoluble, are designed to provide exclusive passage for organics (Table 5.3).

Driving force. – The relevant contributions to the free energy as driving force for membrane transport are the *pressure difference* between liquid feed and vaporized permeate and the difference in *partial pressures* of the permeating species, Eq. 5.2. Of these, the external pressure gradient is of little concern as long as ambient feed pressure is maintained, the difference between "ambient" (feed) and "vacuum" (permeate) amounting to 1 bar at most, Sect. 5.5. The partial pressure of the permeant species in liquid feed and vaporized permeate is given, respectively, by *Raoult's law* (Eq. 2.1) and *Dalton's law* (Eq. 2.4), – presuming that the downstream pressure is sufficiently low to allow the permeate vapor to be treated as a "permanent" gas. The following scheme summarizes the condition:

Feed (l)		**Permeate (v)**	
$p'_i = x_i \gamma_i p^\circ_i$	$>$	$p''_i = y_i p''$	(5.3)

(x_i = mol fraction of target component in feed solution; y_i = same in permeate vapor). It is noted that the phase change, which commonly is assumed to be localized at the downstream interface of the membrane, as in ordinary distillation requires heat of evaporation. That heat is replenished from an external heat source via the thermostated feed stream as indicated in Fig. 5.1.

The *master flux equation* (Sect. 2.3) when incorporating partial pressures as driving force takes this form:

$$J_i = \frac{c^m_i D_i}{z} \ln\left(\frac{x_i \gamma_i p^\circ_i}{y_i p''}\right)$$

or, on rearrangement,   (5.4)

$$J_i = \frac{c^m_i D_i}{z}\left(\ln\frac{\gamma_i p^\circ_i}{p''} - \ln\frac{y_i}{x_i}\right)$$

valid with attention to the boundary conditions concerning membrane permeability outlined above. Flux is seen to increase with increasing activity of the liquid feed component (presuming positively nonideal solution behavior, $\gamma_i > 1$), as well as with decreasing total (gaseous) permeate pressure (p''). It is noted that the permeate pressure is technically influenced by the rate of vapor transfer from membrane to condenser, that rate thus constituting an influence parameter in practical pervaporation.

In single component (pure liquid) pervaporation, $\gamma = 1$ and the mol fractions in feed and permeate each are unity, reducing the driving potential to the difference between saturation vapor pressure (liquid) and permeate pressure (gas), $\Delta p = p_i^\circ - p''$.

Selectivity. – As a measure of the "success" of the separation operation, recording selectivity relies on a comparison of the analytical compositions of feed and permeate, – both changing systematically in the process: As batch operation on a time scale (as in Fig. 5.1), or along the extent of a membrane separation stage (as in Fig. 3.5).

Observed selectivity is influenced by process dynamics. In pervaporation it is concentration polarization and the swelling state of the membrane affecting transport rates. It is therefore necessary to distinguish between the "intrinsic" selectivity of the barrier, and its (lesser) performance under operating conditions. As in cross flow membrane filtration, the feed mixture outside the reach of process dynamics is referred to as *bulk*, sometimes with the attribute "well mixed". Several ways of expressing practical selectivity are in use (i, j = liquid mixture components):

- The first compares directly with binary distillation, accounting for the separation effect in terms of a *separation factor* α_{ij} as depicted by a *McCabe-Thiele diagram* of vapor composition versus liquid feed composition (i = faster moving component).

$$\alpha_{ij} = \frac{(c_i/c_j)''}{(c_i/c_j)'} = \frac{c_i''(1-c_i')}{c_i'(1-c_i'')} = \frac{(p_i/p_j)''}{(c_i/c_j)'} \tag{5.5}$$

As a fitting example, Fig. 5.2 compares evaporation (at reduced pressure) and pervaporation (through a hydrophilic CA membrane) of water-ethanol using water fractions as coordinates. It is observed that hydrophilic pervaporation favors the higher boiling water, enrichment being particularly effective near the azeotropic composition of the feed at about 4 w-% H_2O.

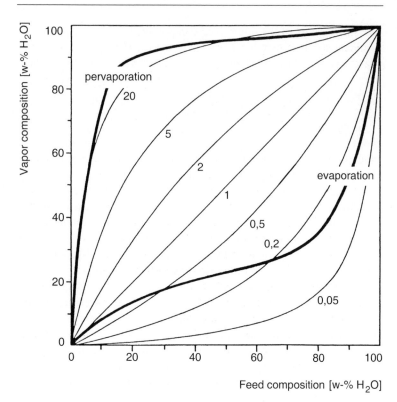

Fig. 5.2. Hydrophilic pervaporation (CA membrane) versus low pressure evaporation of water-ethanol in the presentation of a McCabe-Thiele diagram. Thin lines are curves of constant separation factor relative to the diagonal at which $\alpha_{ij} = 1$ (no separation).

- A practical measure of selectivity takes heed of the fact that, as a rule, pervaporation is applied to separate (enrich) minority components out of dilute feed solutions. If that component is the wanted species, its *enrichment* is of interest, expressed as ratio of concentrations (in weight or volume or molar units) in permeate over feed: The *enrichment factor* $\beta_i = c_i'' / c_i' > 1$, which relates to the *solute rejection R* of reverse osmosis (Eq. 3.7), – enrichment seen as negative rejection (i = minority species):

$$\alpha_{ij} = \frac{\beta_i}{\beta_j} = \frac{(c_i''/c_i')}{(c_j''/c_j')} \qquad \text{and} \qquad \beta_i = 1 - R \qquad (5.6)$$

With some arithmetic ado, separation factor and enrichment factor are interconvertible; at very low feed concentration (at the far ends of the composition range) they come close, $x_i \to 0$ rendering $\alpha_{ij} \to \beta_i$.

- Enrichment of high boiling/low solubility organics from water by organophilic pervaporation easily extends beyond the miscibility limit of the components, leading to phase separation ("demixing") of the permeate once it becomes liquefied. The permeate now consists of two coexisting liquid phases of different composition and density: One an aqueous solution saturated with (little) organic, the other an organic solution saturated with (little) water. "Natural enrichment" under conditions of phase separation is the proportion of the organic component in the organic-rich phase. It is temperature dependent, just as the mutual solubility of the components is (Sect. 5.5 has examples).

- If the minority feed component is unwanted, to be removed in refining a wanted product, loss of product into the permeate is a realistic indication of selectivity. A relevant example is residual alcohol in the aqueous permeate when dehydrating alcohol-water mixtures (Sect. 5.4.1).

5.3   The capability of pervaporation

Figure 5.2 is an example of barrier interference. What a McCabe-Thiele diagram does not show is rates: Pervaporation is "slow" in comparison to "instantaneous" evaporation, raising the question under which conditions the separation capability of pervaporation may be used to advantage. With a view at aqueous-organic liquid separations, these are the facets to be considered:

- *Economy of affinity.* Prevailing modes of liquid-polymer interaction dictate two directions of process design: Selective water permeation through *hydrophilic* barriers, or selective permeation of organics using *organophilic* barriers. By the same argument, using membranes to separate equimolar ("even") liquid aqueous-organic mixtures is an unlikely proposition.

- *Economy of mass transport.* Extending the argument, mass transport economics suggests applying pervaporation to dilute feed solutions, treating the respective minority component to be the preferentially permeating one.
- *Nonequilibrium separations.* Preferential sorption is the key to uncommon separation effects: Selective transport of water by glassy barriers to separate constant boiling mixtures (azeotropes); enrichment, aiming at recovery, of high boiling organics by rubbery polymers ("high boiler pervaporation").
- *Bioseparations.* Membranes allow for gentle (low temperature) and chemically noninterfering liquid separations such as are desirable in bioprocessing; direct coupling to "life" fermenters is a distinct possibility (membrane bioreactor, Sect. 4.5).

In the following, the two principal directions of aqeous-organic pervaporative separation – hydrophilic and organophilic – are presented and illustrated by case studies.

5.4   Hydrophilic pervaporation

5.4.1   General observations

The aim is to dehydrate organic solvents with particular attention to dewatering azeotropes (constant boiling mixtures). Process design is dictated by the predicament of all barriers: They are "leaky". Even the most water-selective membrane will not completely block the passage of organics, turning the nonvalue permeate (water) into a problem waste. Dense hydrophilic membranes, insensitive to organic attack, are thus called for. To offset the limited permeability (low flux) inherent to dense membranes, thin (asymmetrically structured) membranes are employed, preferably in combination with elevated temperatures. As a semi-standard in hydrophilic pervaporation have emerged composite membranes of crosslinked polyvinylalcohol (PVAL) on a microporous support, the thickness of the active layer being in the micrometer range.

The conceptual answer to limited water permeability is *vapor permeation* through microporous inorganic barriers, raising the feed temperature to above boiling, thereby enhancing the rate of mass transport and eliminating the phase change associated with liquid pervaporation. The situation is illustrated in Fig. 5.3, comparing pervaporative flux as function of water content of isopropanol-water mixtures of a polymer membrane (PVAL) and a zeolite barrier (NaA) at the temperatures indicated (permeate pressure $p'' \approx 20\,mbar$).

This is the information conveyed.

Performance. – Pervaporation is "slow" by comparison. For perspective, even at a temperature of 90°C the flux of the PVAL composite membrane at azeotropic feed composition (approximately

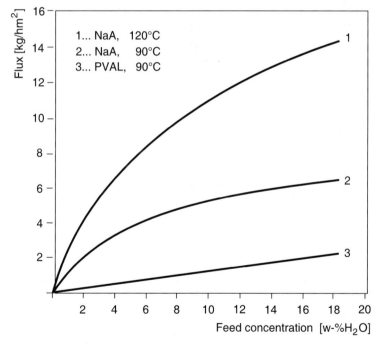

Fig. 5.3. Dehydration of isopropanol by pervaporation and vapor permeation: Flux of a polymer membrane (PVAL composite) and a zeolite barrier (NaA) as function of feed composition at the temperatures indicated [1]. Prevailing curvature reflects Langmuir sorption isotherms.

10 w-% H_2O) is one tenth of what is considered to be an economic lower limit of water flux in reverse osmosis: 40 against 400 L/d m². The superior performance of the zeolite membrane is apparent, especially at the highest temperature shown (120°C), which is likely beyond the long-term reach of most polymeric membranes. Selectivity is judged by the amount of carried-over isopropanol in the aqueous permeate (reported at below 1% in the example shown).

Transport modeling. – As is the design purpose, the process operates against a diminishing water content of the feed (reading the curves from right to left). This brings to focus the concentration dependence of the flux of the target species, designed to be the preferentially permeating minority component of the feed solution. That dependence is related to shape and slope of the sorption isotherm of the permeant-membrane system considered, identifying sorption to be a key transport parameter in pervaporation (c_i^m in Eq. 5.4). At sufficiently low sorption, that is to say, as long as the sorption isotherms are linear or of the *Henry* type, flux varies linearly with feed concentration. The system H_2O-PVAL in Fig. 5.3 shows such a linear dependence throughout, indicating "well behaved" sorption in accordance with a regular (Nernst type) distribution of water between isopropanol and membrane.

Permeance. – A linear dependence of flux on feed concentration conceptually corresponds to a *permeance* describing the proportionality between flux and driving force at given membrane and given operating conditions (Sect. 1.6). Deviations from linearity – now seen as deviation from *well-behaved permeance* – occur in both directions, as documented in Figs. 5.3 (for hydrophilic) and 5.6 (for organophilic pervaporation). The flux curves for the zeolite barrier, when likened to sorption isotherms, resemble the *dual sorption* isotherms known in gas sorption, suggesting pervaporation respectively vapor permeation across non-swelling barriers to be variants of gas permeation. By contrast, pervaporation of organic solutes through high-swelling ("rubbery") membranes shows a higher than linear dependence of organic flux on feed concentration akin to *Flory-Huggins* isotherms, ascribed to a plasticizing effect which simultaneously enhances sorption capacity (polymer swelling) and overall permeant mobility (Sect. 2.3.1).

5.4.2   Pervaporation versus reverse osmosis

Hydrophilic pervaporation and reverse osmosis are related membrane processes in that both are designed to preferentially transport water using basically similar polymeric membranes, both interpreting water transport by a solution-diffusion mechanism (taking exception to inorganic barriers). Conceptual differences notwithstanding, it is instructive to compare the driving forces and resultant fluxes for the two processes.

For a model comparison, pure water transport is considered, identifying the driving force as gradient of water activity between feed and permeate interfaces *inside* the membrane [$a_i'(m)$ and $a_i''(m)$ = boundary activities *within* the membrane]. The argument is based on the swelling state of the membrane under process conditions as it influences permeant activity. Viewing water as incompressible, the activity of the liquid feed is unity for both processes, $a_i' = 1$.

In *pervaporation* there is equilibrium swelling at the feed side of the membrane [$a_i' = a_i'$ (m)], declining to a state of near-zero swelling ("dryness") at the permeate side [a_i'' $(m) < a_i'$ (m)]. The gradient inbetween is pictured naively as linear, as provides for *Fickian* diffusion. (It is noted that the swelling profile in organophilic pervaporation is decidedly nonlinear, Fig. 2.2). The permeate is water vapor, its activity defined as ratio of downstream pressure over saturation (pure component) vapor pressure of water, $a_i'' = p''/p_i^\circ$, ranging from $1 \to 0$ as the downstream pressure is lowered. At equilibrium [a_i'' $(m) = a_i''$] and the relevant activity gradient is that between *liquid feed* and *gaseous permeate*, namely [$1 - a_i''$ (m)].

In *reverse osmosis* the membrane is considered uniformly swollen throughout (*isotropic swelling*), implying level water activity from feed side to permeate side within the membrane, [a_i' $(m) = a_i''$ (m)]. Pressurizing the feed and, in turn, the membrane will lower the water activity inside the membrane by compressing the polymer matrix; water is literally "squeezed out", flowing off freely. The relevant activity gradient, therefore, is that between *liquid feed* and *compacted membrane*, once again [$1 - a_i''$ (m)].

Minimizing the downstream boundary activity of the permeant, a_i'' (m), is seen to be the common handle to enhance the driving potential. Eqs. 5.7 relate the boundary activity to the forces actually employed: Lowering the pressure of the gaseous permeate in pervaporation (p'' $(v) \to min$); raising the hydraulic feed pressure in reverse osmosis (p' $(l) \to max$).

Pervaporation $$a_i''(m) = \frac{p''}{p_i^\circ}$$

(5.7)

Reverse osmosis $$a_i''(m) = exp\left(-\frac{V_i}{RT}\Delta p\right)$$

When comparing the driving forces to effect equal water flux, a surprisingly large advantage in favor of pervaporation is suggested. For example: Theoretically, a reverse osmosis *feed pressure* (p') of 400 bar is needed to produce the same water flux as a pervaporation *permeate pressure* (p'') of 5 mbar (one fifth of the saturation vapor pressure of water, p''/p°) will. Actually, observed pervaporation fluxes consistently are lower than reverse osmosis fluxes. Analysis of this seeming anomaly is a lesson in barrier interference. Three influence factors are held responsible:

- Trivia first: *Membrane thickness.* In form of the active layer of flat-sheet composite membranes, the thickness of current *hydrophilic* pervaporation membranes (PVAL) is higher by an order of magnitude than that of composite reverse osmosis membranes (PA), –2 μm as against 0.2 μm for illustration; water permeance is consequently lower.
- *Membrane swelling* in pervaporation declines from "fully swollen" at the feed side to "virtually dry" at the permeate side, the swelling (= sorption) profile inbetween being a function of the intensity of molecular interaction between permeant(s) and polymer. Irrespective of the shape of the sorption profile, overall water sorption (c_i^m) is lowered and water mobility hindered by regressive sorption, resulting in a transport resistance higher than would be encountered under level swelling conditions, as in reverse osmosis.
- *Structural pressure loss.* Asymmetric (thin) membranes are mechanically stabilized by a microporous support, onto which the membrane is applied skin-like (composite membrane structure). Evaporation of the permeants occurs into the supporting sub-

structure, the associated volume increase, being inversely proportional to pressure, reducing the available driving potential ($p_{eff} > p''$).

5.5 Organophilic pervaporation

5.5.1 General observations

While water is the single target permeant in hydrophilic pervaporation, organophilic pervaporation is as diverse as there are volatile organic solutes (VOC's), – to be recovered (as product) or removed (as contaminant) from aqueous solutions. If biosynthesis is to be a guide, an upper limit to organic concentration encountered is of the order of 10%, as witnessed by the prevalent ethanol concentration in wine; lowest concentration is in the *ppm* range, found, for example, with aroma compounds or else trace industrial water pollutants.

The membranes of choice are rubbery (elastomeric) polymers disposed to swelling when exposed to organic solvents. The prototype of a rubbery hydrophobic polymer is *silicone rubber* (polydimethylsiloxane, PDMS), available in sheet form since 1951, first discovered by Kammermeyer for its extraordinary permeability to gases (oxygen). Other noteworthy elastomers are polyetherblockamide (PEBA) and polyurethane (PUR), both "segment-elastomeric" polymers (Sect. 6.2).

Performance. – All polymer films, even those dubbed "hydrophobic", are permeable to water to a degree, still more so when swollen. As a consequence, while *hydrophilic* pervaporation may strife for exclusive water selectivity, *organophilic* pervaporation enriches the organic solute against an unavoidable undercurrent of water. Concomitant water flux, in turn, is an indicatior for the swelling state of a rubbery membrane: Enhancement of permeability due to swelling is immanent when, as function of organic feed concentration, water flux starts to increase (exemplified by the shaded area in Fig. 5.6). The separation effect under these conditions is determined by comparing the organic target concentration in two aqueous process solutions: One the permeate (following condensation), the other the bulk feed; the ratio of the two is the organic enrichment ($\beta_i = c_i''c_i' > 1$).

Activity coefficients. – From the statement of driving force (Eq. 5.3) organic enrichment in terms of mol fractions is obtained as

$$\beta_i = \frac{y_i}{x_i} \leq \frac{\gamma_i p_i^\circ}{p''} \tag{5.8}$$

Solute enrichment may thus be estimated from the thermodynamic condition of the feed (Henry coefficient, $\gamma_i p_i^\circ$) and from process conditions (downstream pressure, p''), without prior knowledge of membrane properties. The key parameter for further assessment is the *activity coefficient* of the organic solute, both in the aqueous feed solution and in the membrane (polymer) phase. Being concentration dependent (see Fig. 2.1), activity coefficients for consistency are recorded at "infinite dilution", that is, at their numerical highest. Excepting the few instances of negative deviation from Raoult's law (notably carboxylic acids), the solutes under consideration form positively nonideal solutions with water, activity coefficients ranging from $\gamma \approx 2$ for methanol (the species "closest" to water) to several 10^4 for nonpolar solutes. The sheer magnitude of this range is noteworthy; it is paralleled by a vast range of nominal selectivities as exemplified by the data of Table 5.1 (observed with an uncommonly thick

Table 5.1. Organophilic pervaporation of aliphatic alcohols: Correlation between aqueous solution activity coefficient and separation factor. Beginning with n-butanol miscibility with water is limited. Derived from [2].

Alcohol	Boiling point (°C)	Activity coefficient (infinite dilution)	Separation factor (α_{ij})
methanol	64.5	2	9
ethanol	78.3	5	17
propanol	97.2	15	67
butanol	117.5	50	74
pentanol	138	200	265
hexanol	157	1000	1050
heptanol	176	3000	1600

PDMS membrane of 200 μm thickness; 25°C.
Feed concentration: 1 vol-%; hexanol 0.5 vol-%; heptanol 0.1 vol-%

membrane). It is recalled that solubility and activity correlate inversely; for sparingly soluble organics in water the activity coefficient is the inverse of molar solubility, and vice versa (Eq. 2.2).

Activity coefficients for the system ethanol-water (completely miscible) are presented in Appendix A; Table 5.1 has activity coefficients for the homologous series of aliphatic alcohols (going far beyond the realm of miscibility); Table 5.2 has activity coefficients for the four isomeric butanols (near the limit of aqueous miscibility).

When contemplating the influence of activity coefficients on solute relocation (transfer) in liquid separations, three practical situations merit attention:

- Relocation into the vapor phase: *Evaporation* is dictated (and predictable) by VLE; the significance of the Henry coefficient ($\gamma_i p_i^\circ$) is to describe solute *volatility* relative to the vapor pressure of the pure component.
- A variant of evaporation is *steam distillation*: Certain high boiling organics are "volatile with steam", meaning that they are volatilized by steam blown into the feed mixture. Given a low level of molecular interaction (as witnessed by low water solubility), the partial pressures of water and organic in the extracted vapor add up to the sum of the pure component vapor pressures. Organic mol fraction in the vapor phase under these conditions is determined by the vapor pressure of the organic species according to $x_{org} \approx p_{org}^\circ / p_{water}^\circ$ (Eq. 2.3). Condensation of the vapor immediately results in phase separation, the coexisting phases being the two components as nearly pure as their limited mutual miscibility will allow them to be. The organic phase, saturated with water, represents the "natural enrichment" of steam distillation.
- Relocation into a polymeric phase (and thence to vapor): *Pervaporation*, dictated by the distribution of organic solute between aqueous feed and polymeric membrane (Eq. 2.18). Distribution in favor of the polymer phase has two consequences, which are counteracting: (a) Organic enrichment is apt to be higher than suggested by volatility: *High boiler pervaporation* (Sect. 5.5.2); (b) a boundary layer depleted of organic solute will develop, reducing solute permeance: *Concentration polarization*.

5.5.2   High boiler pervaporation

What is intriguing about organophilic pervaporation: It enables aqueous-organic separations which seemingly defy the limitation of volatility of the organic species, even at gentle temperatures. High activity coefficients (low solubility) of the organic solutes in water, combined with preferential sorption by the membrane polymer, result in high (occasionally extreme) enrichment of low volatile organics from aqueous solution. The case studies below illustrate some aspects of *high boiler pervaporation*.

Mostly water, after all. – High enrichment notwithstanding, most of the permeate still will be water. The following example is taken from the repertoire of microbially accessible (and thereby "natural") aroma compounds. *γ-Decalactone* (MW 170.2; bp. 281°C) is an aroma compound of "fruity" fragrance (peach), forming a highly nonideal solution with water (activity coefficient ~ 14000). Vacuum distillation (VLE) of the aqueous solution has but little effect on composition, a high activity coefficient just about offsetting the low vapor pressure ($\gamma_i p_i^\circ \approx 1$). Pervaporation, on the other hand, enriches γ-decalactone from a feed concentration of 100 ppm (0.01%) to a permeate concentration of nearly 3% (PEBA; 40°C). The necessary information is that even at a 300-fold enrichment of the organic target species, 97% of the permeate still is water. Organic flux, for perspective, is low: 0.8 g/h m² of γ-decalactone in the example presented.

Phase separation. – High boiler enrichment, as a rule, leads to phase separation (demixing) of the condensed permeate in accordance with the *phase diagram* (diagram of miscibility versus temperature) of the aqueous-organic system under consideration. Phase separation may be employed as a means to improve the overall selectivity of high boiler pervaporation.

Phenol (MW 94.1; bp. 182°C) is a rewarding study object, first of all for its eminent industrial relevance and water polluting prowess, but also on account of its relatively high water solubility thanks to the weakly acidic function of the OH-group. At a temperature of 30°C phase separation commences at a phenol concentration of about 10 w-%, yielding two coexisting liquid phases,

- "phenol in water", approximately 10% phenol; and
- "water in phenol", approximately 70% phenol,

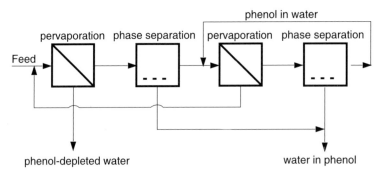

Fig. 5.4. Two-stage pervaporation/phase separation process scheme for phenol recovery at the enrichment level of "water in phenol" [3].

the phenol-rich phase precipitating due to higher density. A process scheme combining pervaporation and phase separation to effect phenol enrichment to the level of "water in phenol" is presented in Fig. 5.4. The permeate of the first pervaporation stage, enriched to a phenol concentration of $>10\%$, undergoes phase separation to yield the two coexisting phenolic solutions indicated. Of these, the fraction "water in phenol" is withdrawn as target product while the fraction "phenol in water" is subjected to a second pervaporation/phase separation stage, once more producing "water in phenol" to be recovered, and "phenol in water" being recycled as shown (a *feed and bleed* situation). Nominal phenol enrichment is the ratio of concentration of the phenol-rich phase (fixed by the phase diagram) to that of the prevailing feed solution.

Salting-out. – Upon addition of salt to aqueous solutions the activity of water is lowered (colligative properties, Sect. 2.1.1), whereas the activity of dissolved nonelectrolytes increases, both effects having the same origin: a decrease in "availability" of water. In keeping with the influence of activity coefficients on pervaporation performance, salting-out is expected to enhance both organic flux and organic enrichment. Confirmation is presented in Fig. 5.5, again using phenol as sample solute.

The effect is clearly confined to nonelectrolytes, as again the pervaporation behavior of phenol demonstrates. With increasing pH value phenol (a weak acid) gradually converts into ionic phenolate: $C_6H_5OH + NaOH \Leftrightarrow [C_6H_5O^-] \, Na^+ + H_2O$. Pervaporative phenol en-

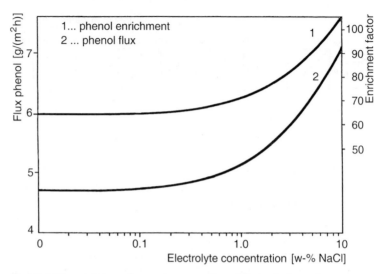

Fig. 5.5. Salting-out to enhance pervaporation: Effect of electrolyte concentration on flux and enrichment of phenol. Feed 400 ppm; PEBA membrane; 50°C [4].

richment as function of pH diminishes precisely as the concentration of undissociated phenol in equilibrium with phenolate does. Conversely, phenolate is rejected by reverse osmosis whereas phenol is not. The cross-over between phenol enrichment and phenolate rejection occurs at a pH value equal to the dissociation constant, $pH = pK_a = 10.4$.

Concentration polarization. – Concentration polarization is a phenomenon of process dynamics. Preferential sorption under conditions of mass transfer will lead to accelerated depletion of organic solute near the feed-membrane interface. The causality of high solute activity coefficient in the feed leading to high solute sorption by the membrane (Eq. 2.18), in turn leading to high solute enrichment on pervaporation suggests a correlation between organic enrichment and concentration polarization.

Estimation of the significance of concentration polarization is based on a comparison of "actual" (experimental) solute enrichment data with the "intrinsic" enrichment of a given membrane. *Actual* enrichment is the ratio of solute concentrations in permeate and bulk feed, and is affected by boundary layer depletion; *intrinsic* enrichment is the highest achievable by the membrane in the absence of boundary layer effects. Evaluation of the *polarization equation* (Eq. 4.4) to emphasize the concentration polarization modulus $c_w/c_b \leq 1$ yields the following expression [5]

$$\frac{c_w}{c_b} = \frac{exp\left(J_v\delta/D\right)}{1+\beta_o\left[exp\left(J_v\delta/D\right)-1\right]} \tag{5.9}$$

showing concentration polarization to be dependent on intrinsic enrichment (β_o) and process conditions through the ratio of convective and diffusive mass transport in the boundary layer ($J_v\delta/D$). Pervaporation being "slow", it is the mass transfer coefficient (Eq. 4.5) and the intrinsic enrichment which are of concern. Surveying available data it appears that concentration polarization in organophilic pervaporation requires attention at solute enrichment higher than $\beta \approx 100$. It is noted that this proviso necessarily limits the feed concentration of concern to below 1%. Primarily affected by concentration polarization, therefore, are sparsely soluble organic solutes such as aroma compounds and higher alcohols (Table 5.1), which may exhibit extreme nominal enrichment.

On the other hand, low MW commodity chemicals of biosynthetic promise, including lower alcohols, are far from the critical limit, pervaporative recovery from the respective fermentation broths proceeding at more modest enrichment: Ethanol by a factor of about 5; n-butanol by a factor of about 30 (PDMS membranes).

5.5.3 Butanol, a glimpse at bioseparations

When Weizmann introduced ABE (acetone-butanol-ethanol) fermentation in 1912, the incentive was *acetone,* – badly needed by the military for the manufacture of smokeless gunpowder according to one of Nobel's countless patents. *Ultrafiltration* using asbestos fiber devices to clarify fermenter broths was known and practiced at the time, however, recovery of volatile metabolites was strictly by distillation from batch fermentation. *Butanol* (n-butanol, Table 5.2), already then recognized as a precursor to synthetic rubber, today would be the product of value.

Today, Weizmann would have considered a membrane bioreactor (MBR; Sect. 4.3), product recovery by pervaporation inclusive. A sampling of butanol pervaporation, laboratory scale, follows.

Table 5.2 Physical constants of the structural isomers of butanol (C$_4$H$_9$OH; MW 74.12).

		Boiling point	Vapor pressure at 20°C	Solubility in water at 20°C	Activity coefficient at 20°C	Water content in azeotrope	
		(°C)	(mbar)	(w-%)		(w-%)	(°C)
n-butanol	●-●-●-●-OH	117.5	5.7	7.7	41.1	42.5	(93)
iso-butanol	●-●-●-OH ●	108	12	8.5	44.4	32	(90)
sec-butanol	● ●---●-OH OH	99.5	16	12.5	20.8	30	(88)
tert-butanol	●-●-OH ● ●	82.5	41	miscible	11.4	11.8	(80)

There are four structural isomers of butyl alcohol (C_4H_9OH; MW 74.1); Table 5.2 summarizes the physical constants. Two of the alcohols are seen to be "high boilers"; water miscibility and aqueous activity correlate inversely, only tert-butanol, closest to water in activity, being completely soluble; all butanols form positive azeotropes with water. Why microbial action exhibits a preference for *n-butanol* (soon to develop "product inhibition") is a matter of speculation.

Limited miscibility leads to phase separation. The coexisting phases at room temperature in case of n-butanol are

- "butanol in water", 7.7% butanol;
- "water in butanol", exceeding 60% butanol,

the latter representing the "natural limit" of enrichment.

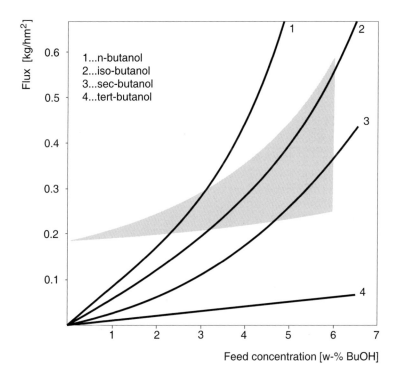

Fig. 5.6. Organophilic pervaporation of the butanol isomers: Organic flux (curves) and realm of concurrent water flux (shaded area) as function of organic feed concentration (PEBA membrane; 50°C) [6]. Prevailing curvature reflects Flory-Huggins sorption isotherms.

Figure 5.6 summarizes the organophilic pervaporation of the butanol isomers from aqueous solution as *flux* versus *feed concentration* (permeance). [Fig. 5.3 is an analogous presentation for hydrophilic pervaporation]. What is observed is an illustration of pervaporation thermodynamics:

- The order of fluxes is counter to the order of both boiling points and solubility, the highest boiling and least soluble n-butanol showing highest permeance.
- Up to an organic feed concentration of ~ 1 w-% (the presumed limit to biotoxicity of n-butanol in ABE fermentation) fluxes increase linearly, the slopes reflecting the order of *Henry* coefficients; thereafter, flux increase is stronger than linear.
- Water flux, high to begin with, is practically independent of organic feed concentration up to the 1% limit, increasing thereafter in compliance with polymer plasticization; true to the general pattern, the effect is least pronounced with tert-butanol.

Table 5.3 compares the performance of three "classical" elastomers, a polymer blend incorporating silicalite (a hydrophobic molecular sieve), and a supported liquid membrane (oleyl alcohol, $C_{18}H_{35}OH$) with equilibrium evaporation in n-butanol enrichment. While organic flux (= flow density) would be a key parameter in establishing the membrane area of a fictitious membrane bioreactor,

Table 5.3. Pervaporation of n-butanol through organophilic membranes and a liquid membrane, comparison with evaporation. Data in brackets estimated from published accounts. Taken from [6] and [7].

Membrane	Flux [g/h m^2]		Flux ratio	Selectivity	
	BuOH (i)	H$_2$O (j)	H$_2$O/BuOH	α_{ij}	β_i
PDMS	26	44	1.7	59	37
PEBA	56	222	4	25	20
PUR	10	78	8	13	11
PDMS-SIL	(90)	(70)	(0.8)	(120)	(55)
Oleyl alcohol (SLM)	(50)	(35)	0.7	150	60
Evaporation (VLE)	-	-	-	25	20

Feed 1 w-% n-BuOH; membrane thickness 50 μm; temperature 50°C.
SIL = silicalite molecular sieve in PDMS (~ 50%), composite on PEI
SLM = supported liquid membrane in microporous PP of 25 μm thickness
VLE = vapor-liquid equilibrium

it is the concurrent water flux which determines the enrichment achieved. Accordingly, highest organic enrichment is observed with the liquid membrane, conceptually having least interaction with water.

5.6 Pervaporation in perspective

All things considered, pervaporation may well be the most versatile and least adopted of barrier separations. A number of recent reviews testify to the continuing fascination [8]–[14].

As an industrial separation process, pervaporation is far from mature. Basically there are two problem areas, unrelated on first sight only:

- Scaling up has not (yet) been achieved;
- Biogeneration of base chemicals is not (yet) attractive.

Scaling up faces the problem of handling (transporting and condensing) large volumes of low pressure vapor. Module design, by adhering to high pressure reverse osmosis prototypes, does not meet pervaporation needs adequately; alternative designs would have to be modeled after the low pressure/high throughput membrane configurations employed in cross flow microfiltration (pleated membrane designs).

The generic "tree" branching into ever higher levels of carbonic chemistry from roots nourished by either *petroleum* or *biomass* is well known. As long as oil is "cheap", prospects of seeing biotechnology developing into a serious contender of petrochemistry are regrettably dim. This goes for the development of the supporting membrane technology as well.

Bibliography

References

[1] T. Melin, R. Rautenbach, S. Sommer, U. Hömmerich, Einsatzpotential von anorganischen Zeolithmembranen in der Pervaporation und Dampfpermeation. Chemie Ingenieur Technik 70 (1998) 1101–1102.
[2] J. M. Watson, P. A. Payne, A study of organic compound pervaporation through silicone rubber. J. Membrane Sci. 49 (1990) 171–205.

[3] K. W. Böddeker, Pervaporation of aqueous phenols. Proc. 2nd Int. Conf. Pervaporation Processes (R. Bakish, ed.), San Antonio, Texas, 1987.

[4] K. W. Böddeker, G. Bengtson, E. Bode, Pervaporation of low volatility aromatics from water. J. Membrane Sci. 53 (1990) 143–158.

[5] R. W. Baker, J. G. Wijmans, A. L. Athayde, R. Daniels, J. H. Ly, M. Le, The effect of concentration polarization on the separation of volatile organic compounds from water by pervaporation. J. Membrane Sci. 137 (1997) 159–172.

[6] K. W. Böddeker, G. Bengtson, H. Pingel, Pervaporation of isomeric butanols. J. Membrane Sci. 54 (1990) 1–12.

[7] J. Huang, M. M. Meagher, Pervaporative recovery of n-butanol from aqueous solutions and ABE fermentation broth using thin-film silicalite-filled silicone composite membranes. J. Membrane Sci. 192 (2001) 231–242. M. Matsumura, H. Kataoka, M. Sueki, K. Araki, Energy saving effect of pervaporation using oleyl alcohol liquid membrane in butanol purification. Bioprocess Engineering 3 (1988) 93–100.

Reviews

[8] X. Feng, R. Y. M. Huang, Liquid Separation by membrane pervaporation: A review. Ind. Eng. Chem. Res. 36 (1997) 1048–1066.

[9] A. Baudot, M. Marin, Pervaporation of aroma compounds: Comparison with vapour-liquid equilibria and engineering aspects of process improvement. Trans. Inst. Chem. Eng. 75 (1997) 117–142.

[10] F. Lipnizki, R. W. Field, P-K. Ten, Pervaporation-based hybrid process: A review of process design, applications and economics. J. Membrane Sci. 153 (1999) 183–210.

[11] T. C. Bowen, R. D. Noble, J. L. Falconer, Fundamentals and applications of pervaporation through zeolite membranes. J. Membrane Sci. 245 (2004) 1–33.

[12] L. M. Vane, A review of pervaporation for product recovery from biomass fermentation processes. J. Chem. Technol. Biotechnol. 80 (2005) 603–629.

[13] C. C. Pereira, C. P. Ribeiro Jr., R. Nobrega, C. P. Borges, Pervaporative recovery of volatile aroma compounds from fruit juices. J. Membrane Sci. 274 (2006) 1–23.

[14] P. Shao, R. Y. M. Huang, Polymeric membrane pervaporation. J. Membrane Sci. 287 (2007) 162–179.

6 What Membranes are About

6.1 Prelude: Collodion membranes

Collodion, the viscous solution of cellulose nitrate in ether-alcohol, is the prototype of "synthetic" film-forming materials and as such is the parent of artificial membranes, likewise of photographic film. Cellulose nitrate is also the parent of artificial fiber (*rayon*) and of plastics (*celluloid*), not to mention explosives (in that capacity better known as *nitrocellulose*). It was discovered by Schönbein (who also discovered ozone) in 1846, who coined the name *guncotton* in hopes of exploiting its hazardous nature. Cotton is cellulose at its natural purest; collodion uses partially nitrated cellulose also known as *pyroxylin*.

Collodion membranes are prepared by allowing the solvents to evaporate from the viscous solution ("dope") spread unto a smooth surface which, in the old days, sometimes was a pool of mercury. Controlled evaporation (first ether, then alcohol) causes the cellulose ester to precipitate into a cohesive film, evaporation conditions permitting to influence the permeability characteristic of the resulting membrane. Next to thickness and morphology, the all-important parameter determining water permeability is the water content of the membrane, which is controlled by immersing the film in water before evaporation of the organic solvents is complete, whereupon the remaining solvents (mostly alcohol at this stage) are being exchanged against water. By this procedure, a membrane water content of from 50 to 90% is attainable, most of which in the form of "pore fluid". This is to be contrasted with hydration following complete solvent evaporation, which, by order of magnitude, amounts to only 10%, and is seen as "structural" (truly absorbed) water. – It is noted that the water content of the dense (nonporous) skin layer of asymmetric reverse osmosis membranes is of the same order (Sect. 1.3), albeit at thickness of 0.2 μm as against 200 μm for the symmetrical collodion film.

The process of membrane formation described, referred to as *phase inversion* technique, in countless variations is still in use today [1], as are microporous cellulose nitrate membranes for use in *ultrafiltration*.

First mention of collodion membranes is by Fick in his classic studies on liquid diffusion (1855) [2]. His membranes were completely dry prior to exposure to water and, consequently, were "tight", showing low water permeability and practically no salt leakage in osmotic experiments. Comparison with the much more "open" animal membranes (pig's bladder) lead him to dismiss physical pores, vaguely speculating instead on "interstitial molecular diffusion" as mechanism for water transport. The long-standing controversy about "pores or no pores" in interpreting membrane permeation has its origin here.

6.2   Membrane polymers – polymer membranes

A list of commonly used (frequently quoted) membrane polymers is presented in Appendix D, arbitrarily arranged in terms of increasing *glass transition temperature* (T_g).

There is no single consistent system or figure of merit by which to categorize membrane polymers. Instead, characterization is by a number of practical criteria such as

- ways and means of polymer formation, chemical and structural identity of polymeric materials;
- film forming properties, manufacturing conditions for polymeric membranes, porous and nonporous;
- barrier properties, performance of polymeric films in fluid transport and fluid separations.

The glass transition temperature used as guiding principle in this survey reflects structure-relevant features of the polymeric materials, foremost *chain flexibility* and *chain interaction*; it is *not* a natural constant. It is based on the descriptive notion that, at high enough temperature, all macromolecular organic structures are somehow mobile and pliable, in short "rubbery". On cooling the random mobility freezes into a "glassy" state at the glass transition temperature, discernible as change in slope of the temperature dependent volume contraction (Fig. 6.1). Polymers with T_g below room temperature (Appendix D), consequently, are considered *elastomers*; all others are *amorphous* or *semicrystalline* with various degrees of crystallinity appearing respectively disappearing at the glass transition temperature. High T_g values indicate high thermal and, implicitely, high chemical stability; the socalled "engineering plastics" are high T_g polymers.

Chemical stability of organic polymers relates to solvent compatibility as a film forming criterion: As a rule not without exception, low T_g membranes are manufactured from suitable polymer solutions (Sect. 6.3), whereas high T_g membranes, being insoluble in common solvents, are formed either by in-situ interfacial polymerization (aromatic PA) or else from melt-extruded film by physical

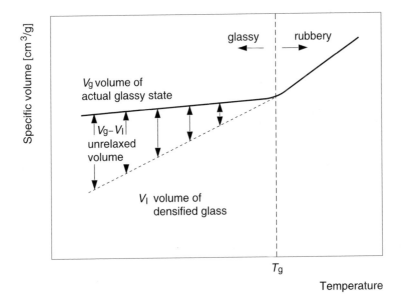

Fig. 6.1. Glass transition: Schematic presentation of the temperature dependence of the specific volume of polymeric materials. Glass transition temperatures are reported from –120°C for PDMS to +300°C for PI (Appendix D).

methods like stretching (PTFE, PP, PE) or track-etching (PE, PC). As a consequence, it is mainly from low T_g polymers that "homogeneous" (nonporous) membranes are accessible, all others variously yielding microporous membrane structures (Sect. 6.4).

Solvent compatibility (chemical stability) has a different connotation when it comes to membrane performance in terms of the two prevailing mechanisms of mass transport, – solution/diffusion and convection/diffusion. In membrane separation of aqueous solution systems a key criterion is the *hydrophilic* versus *hydrophobic* character of the membrane polymer or membrane surface. Hydrophilic membranes are in demand; in Appendix D, hydrophilic polymers are identified by *italics*. While water generally is a nonsolvent for polymers (excepting water-soluble specialty polymers included as "synthetic macromolecules" in Appendix C), water sorption is a condition for osmotic as well as pervaporative water transport; further, the only viable safeguard against protein fouling in (subcritical) ultrafiltration appears to be the hydrophilicity of the membrane respectively membrane surface.

Chain propagation as mechanism of formation of high polymers is not limited to homopolymers: Copolymerization and blending of different monomers offer ways to "design" wanted membrane polymeric materials. Some design principles are typified as follows:

- *Random copolymers* of low T_g moieties, such as the synthetic rubbers listed in Appendix D.
- *Block-copolymers*, composed of soft (low T_g) and hard (high T_g) segments in (more or less) stoichiometric order, the series of polyetherblockamides (PEBA) as example.
- *Polymer blends*: high T_g polymers are hydrophilized by blending with hydrophilic polymers, foremost polyvinylpyrrolidone (PVP); Sect. 4.5 has an example.
- *Filled membranes* are polymer blends incorporating inorganic fillers such as zeolites. In the extreme, the role of the polymer reduces to that of a binder; an example is silicalite in PDMS, Sect. 5.5.3.

Cellulose and cellulose derivatives are a class of film forming polymers by themselves, too well documented to be reiterated [3].

6.3   Like dissolves like

To no small degree, polymer-solvent interaction is at the core of liquid barrier separations, –

- as necessary condition for mass transfer according to the solution-diffusion mechanism, – the sorption contribution (Sect. 2.3.1);
- as dissolved polymer in preparation for membrane manufacture by any of various solution casting procedures (Sect. 6.1).

As solution systems polymer-solvent solutions are dilute and highly nonideal: Limited sorption of solvent (as permeant) in a swollen polymer phase in the first instance; limited dissolution of polymers in organic solvent systems to form viscous *casting solutions* in the second.

Rationalizing *polymer-solvent compatibility* is the same in both instances, and is based on the independent premise that a correlation exists between the *cohesive energy* (potential energy) of pure substances and their mutual miscibility. The cohesive energy of a pure substance (solid or liquid) is the sum total of inter-molecular forces which define the condensed state, and which need to be overcome on *evaporation*. In decreasing order the forces of concern are hydrogen bonds ($\delta_h \sim 40$ kJ/mol), polarity interaction ($\delta_p \sim 20$ kJ/mol), and dispersion forces ($\delta_d \sim 2$ kJ/mol). Genuine chemical bonds (~ 400 kJ/mol) are not affected by evaporation.

The *cohesive energy density* (CED) is the heat of evaporation at constant molar volume, by which the *solubility parameter* δ_i of component i is formally defined (Hildebrand, 1916):

$$\delta_i = \left(\frac{\Delta E}{V}\right)^{1/2} \quad \text{and} \quad \delta_{total} = \left(\delta_h^2 + \delta_p^2 + \delta_d^2\right)^{1/2} \qquad (6.1)$$

In order for two volatile substances to be compatible their free energy of mixing (as heat of mixing, ΔH) is supposed to be small (Sect. 2.1.2). The concept of *solubility parameters* seeks to predict mixture compatibility by relating the heats of evaporation of the mixture components to the heat of mixing: ΔH is small when the difference in solubility parameters is small; highest compatibility (miscibility) is therefore expected at $\delta_i \approx \delta_j$. This is the statement of "like dissolves like" in terms of solubility parameters.

For practical use, total and partial solubility parameters are tabulated in units of $(cal/cm^3)^{1/2}$, the format apparently chosen to gain tractable numbers [4]. Nonvolatile polymers are assigned solubility parameters according to the like-dissolves-like principle by probing their solubility in solvents of known solubility prowess. A practical tool to illustrate polymer-solvent interaction is a *solubility map*, a two dimensional graph which uses the prevalent interaction paramters δ_h and δ_p as coordinates (Fig. 6.2). On this graph every solvent of interest is positioned according to its characteristic solubility parameter listing (in convenient units between 0 and 16), ranging from *hexane* (lower left: no hydrogen bonds, no polarity) to *water* (upper right: maximal on both counts). Any given polymer, if at all responsive to solvents, is inscribed into the solubility parameter map

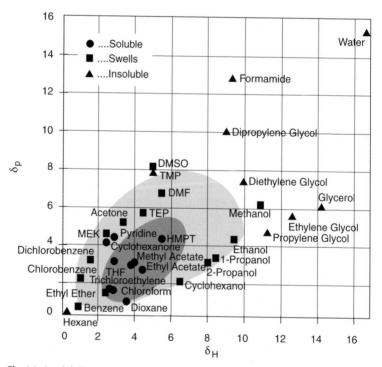

Fig. 6.2. A solubility parameter map for ethyl cellulose. The coordinates represent the principal forces which determine the cohesive energy of the solvents indicated: Hydrogen bonds (δ_h) and polarity interaction (δ_p). After [5].

according to the criteria *soluble, swelling, insoluble,* graphically forming a "compatibility island" with blurred contour.

Aside from examining individual polymer-solvent compatibility, the real value of the solubility parameter analysis is that it allows to compose mixed solvents from nonsolvents or solvents/nonsolvents. The guiding principle is the condition that the tie line between nonsolvents on the map meets the "island" of polymer-solvent compatibility. As an example for the power of mixed solvents, it is noted that cellulose nitrate is insoluble in either alcohol or ether alone, but is soluble in a mixture of the two (collodion, Sect. 6.1). Experimenting with the counteracting influences of solvents and nonsolvents produces phase-inversion membranes, the nonsolvent as a rule being water.

6.4 Microporous barriers

Microporous membranes represent the filtration aspect of barrier separation as opposed to the solution-diffusion behavior of "homogeneous" membranes. As such, microporous membranes are basically governed by size considerations, retaining their structural identity in the filtration operation, whereas homogeneous membranes rely on specific polymer-permeant interactions which typically cause the membrane to swell.

Within membrane separation technology, microporous structures serve a variety of purposes determining material selection, target structure, and method of preparation. Typical uses are:

- Their immediate use as barriers in ultra- and microfiltration in one of the prevailing configurations of *planar* (spiral wound; sandwich; pleated), *cylindrical* (tubular), or *hollow fiber* (microtubular);
- their application as non-wettable porous barrier in membrane distillation respectively osmotic distillation;
- their use as support for immobilized liquid membranes (SLM), or else for enzymes (catalysts) in membrane bioreactors (MBR);
- their application as rigid porous barriers in gaseous diffusion and aerosol filtration;
- their use as structural support in composite membrane constructs, as by coating, interfacial polymerization, or dynamic precipitation.

A pictorial record of microporous structures along with some indication of the techniques employed to create them is presented in Appendix E, drawing on information provided by the manufacturers identified, – expanding between the infinite variety of the *silica remains* of ancient aquatic algae (kieselguhr, E.1) and the exacting layers of the *protein remains* of processed archeo-bacteria (bacterial S-layers, E.18). Clearly, electron microscopy is an indispensable tool in elucidating membrane structure, limited only by the requirement that specimen need to be "dry"; water containing polymer membranes formed by solvent-nonsolvent interplay thus need to undergo a solvent exchange treatment before being ready to reveal their structure.

Membranes of uniform pore structure are termed *isoporous*. Closest to isoporosity among the structures shown are the bacterial S-layers (E.18) followed by track-etched porous films (E.9) and honeycombed alumina (E.14). Woven fabric, polymeric or metallic, of uniform mesh (not shown) is isoporous.

Where to from here? Life's functioning is unthinkable without membranes. *Biomimicry,* – learning how nature does –, is bound to have an influence on future membrane science. Short of living membranes, even life's material science still needs elucidating [6]: How do shells grow (biomineralization)? How to reproduce spider web and silk (noncellulosic natural fibers)?

Bibliography

[1] R. E. Kesting, A. K. Fritzsche: Polymeric Gas Separation Membranes. Wiley-Interscience, New York etc., 1993.
[2] A. Fick, loc. cit. Chap. 7, Ref. [8].
[3] A.F. Turback (ed.): Synthetic Membranes. Volume 1: Desalination; Volume 2: Hyper- and Ultrafiltration Uses. ACS Symposium Series 153 and 154, Washington, 1981.
[4] A. F. M. Barton: Handbook of Solubility Parameters and Cohesion Parameters. CRC Press, Boca Raton, Florida, 1983.
[5] E. Klein, J. K. Smith, Asymmetric membrane formation, solubility parameter for solvent selection. Ind. Eng. Chem., Prod. Res. Dev. 11 (1972) 207–210.
[6] J.M. Benyus: Biomimicry, Innovation Inspired by Nature. HarperCollins Publishers, New York, 1997.

7   Tracing Membrane Science, an Historical Account

First published in Journal of Membrane Science, 100 (1995) 65–68.
Semipermeability of an animal membrane was discovered by chance, if not by accident, by *Nollet* in 1748 [1]. This in short is the story: To prove that it is dissolved air which causes ebullition of liquids under reduced pressure, he intended to store a sample of deaerated alcohol under conditions which would preclude any contact with air for repeat experiments. To this end he closed the filled sample vial with a piece of pig bladder, much as is done today with flexible sealing film, and submerged it in water. The inevitable happened: water being drawn into the alcohol, thereby straining the membrane. Fascinated by the phenomenon, *Nollet* devised some clarifying experiments which established the preferential permeability of his membrane towards water. All this is recorded as an addendum to a treatise on the ebullition of liquids, a topic quite unrelated to membranes [1].

The force or "new power" manifested by the strained membrane remained mysterious until *Gibbs* consolidated the free energy concept in 1873. *Nollet* himself, in a later textbook on experimental physics, relates the effect to the volume reduction observed when alcohol and water are mixed as indicating a natural tendency for interpenetration of miscible liquids [L'art des expériences, ou avis aux amateurs de la physique, par M. l'Abbé Nollet, seconde édition, tôme troisième. Durand, Paris 1770, p. 104].

To *Dutrochet* [2], who introduced the term osmosis to spontaneous liquid flow across permeable partitions, the cause is electricity, "although I admit that I did not succeed in obtaining a reading on the galvanometer, even after several attempts". Noting that liquid flow occurs both ways (exosmosis and endosmosis in his parlance), he disproves an earlier capillary theory by *Poisson*: Capillary action (another force to which considerable attention was paid at the time)

Fig. 7.1. Maurice Quentin de la Tour: Abbé Jean Antoine Nollet (1700–1770). Munich, Alte Pinakothek. Reproduced by permission.

would predict flow exclusively in favor of that liquid which rises highest in a vertical capillary. – As an aside it may be mentioned that *Dutrochet* outpaced many a future membranologist in adopting the metric system.

Graham is best remembered for his contributions on gas permeation, one of his earliest communications being the "notice of the singular inflation of a bladder" of 1829 [3]. A moist bladder partially filled with air or methane (coal gas), when immersed into an atmosphere of carbon dioxide, becomes inflated. Once more, Nature's powerful urge to equilibrate, made visible through the interference of a membrane. For it is not the carbon dioxide moving in, but air barred from moving out, which causes the bladder to bulge. The real significance of this little note, which also appeared in Schweigger-

Seidel's Jahrbuch der Chemie und Physik for that year ["Notiz über das sonderbare Anschwellen einer Thierblase", Band III (1829), pp. 227–229], is that it contains the original statement of the solution-diffusion concept, reiterated more extensively in 1866 [7].

While animal membranes are microporous and hydrophilic, rubbery membranes of plant origin (gum elastic; caoutchouc) are homogeneous and hydrophobic (*Mitchell*, 1829 [4]). With only these two types of membrane available to him, *Mitchell* finds the ratio of permeation rates of various gases to be independent of the membrane used, whereas liquid permeation depends on both the nature of the liquids and the membrane. Selective withdrawal of oxygen from air through a gum elastic membrane into water makes him think of a method to obtain nitrogen gas (the reverse situation would later be known as the artificial gill). Speculating on the power of penetration, a resemblance is drawn between sorption affinity and the ease of condensation of certain gases by charcoal. Sorption (e. g., of carbon dioxide by gum elastic) is envisioned as "interstitial infiltration", leaving no room for conducting capillaries. Indeed, *Mitchell's* acidly polite dismissal of *Graham's* notion of capillary canals, besides making for amusing reading, foreshadows the controversy between those advocating a solution-diffusion mechanism and those insisting on pores when describing membrane permeability, – irrelevant at last with the advent of molecular modeling.

A matter of real confusion is *Graham's* law: Two "square-root laws" of gas transport go by his name, which, in practically identical terms (fluxes being in the inverse ratio of the square roots of their molecular masses), describe two entirely different rate processes.

One is *Graham's* Law of Diffusion, published in 1833 [5], applying to the interdiffusion of gases at uniform pressure. Intermolecular collisions are essential to this process, such as would result in hydrodynamic or viscous flow when proceding through pores. Unidirectional viscous flow, of course, termed transpiration in the early literature, does no longer obey *Graham's* law but is described as *Poisseuille* flow; it is obviously to no separative effect.

The other is *Graham's* Law of Effusion, published in 1846 [Phil. Trans. Royal Soc. (London) 4 (1846) 573], applying to the rates of effusion of gases through small apertures into a vacuum. In true effusion, intermolecular collisions are insignificant, the gas molecules crossing the barrier independently of one another. Gas flow in this limiting situation, originally termed atmolysis, is known as molecular flow or *Knudsen* flow, and yields separation effects as predicted by a square-root law.

In convenient generalization, any gas separation effected by means of a porous barrier is nowadays considered a case of *Graham's* law; the process itself is "gaseous diffusion".

It was *Knudsen* who, much later, recognized the geometrical aspect of it all [6]. The *Knudsen* number (named after him, not by him) relates the mean free path of the gas molecules (primarily a function of pressure) to the dimensions of the duct (diameter and length). It numerically identifies the flow regime between low (viscous flow = transpiration) and high (molecular flow = atmolysis) over three orders of magnitude. As to the interaction of gas molecules with a solid wall, the essential feature of the *Knudsen* theory of gas flow is that the direction into which an impacting molecule is repelled is independent of the direction of impact, the analogy being that of a glowing wall.

Graham's paper "on the absorption and dialytic separation of gases by colloid septa" of 1866 [7] is usually quoted to be the foundation of the solution-diffusion model of membrane transport, exemplified at the fractionation of air through a rubbery membrane. Whereas gaseous diffusion based on molecular flow through pores would slightly favor the lighter nitrogen, the rubber membrane is found to enrich oxygen. Drawing on the solubility of air in water (a subject which occupied *Nollet* [1]), dissolved gases are considered liquefied and thereby amenable to liquid diffusion. A correlation is consequently expected between the penetration of rubber by different gases and their ease of liquefaction, as was already noted by *Mitchell* [4], however, relative rates of solution-diffusion do not yield to rationalization as "squarely" as do those of gaseous diffusion.

Back to liquids. It remained to *Fick* (1855) to interpret the liquid analog to gaseous interdiffusion: The law by which a solute dissipates in its own solvent [8]. The experimental data from which *Fick* set out are again *Graham's*. The idea, which presented itself "quite naturally", was to draw a parallel to the diffusion of heat (*Fourier*) and that of electricity (*Ohm*) in their respective conductors, – no small feat for a demonstrator of anatomy that he was at the time. *Fick's* attempts to model the diffusion of salt solutions through porous partitions appear somewhat tedious, however, they produced the first mention of a collodion membrane, exhibiting an "endosmotic equivalent" (ratio of water to salt diffusing) vastly higher than that of animal membranes.

As to *van't Hoff's* lasting contribution to solution theory [9], such is the beauty of the apparent analogy between gas pressure and osmotic pressure of dilute solutions that it persists although proven wrong a long time ago. It has produced, nevertheless: The idea of the semipermeable membrane; a reminder that molarity is the basis to compare mass action in chemistry; an indication, privately hinted at by *Arrhenius*, that if molarity does not work it might be due to electrolytic dissociation; the concept of isotonic solutions having equal vapor pressure of the same solvent; and the first *Nobel* prize in chemistry (1901). Actually, it was *Pfeffer* who lead *van't Hoff* on the false track with his measurements of osmotic pressure of sugar solutions [W. Pfeffer: Osmotische Untersuchungen. Studien zur Zellmechanik. Verlag W. Engelmann, Leipzig, 1877]. Indeed, the numerical correspondence of the osmotic pressure of an aqueous sugar solution with the ideal gas pressure on an equimolar footing must be regarded as one of nature's profound jokes.

As a preliminary study to an investigation of the state of soaps in aqueous solution, *Donnan* examined solutions of Congo red, it being known that this "colloidal" sodium salt will dissociate but not diffuse through parchment paper. When such a solution is contacted, across a parchment paper diaphragm, with a sodium chloride solution, all ions present except the bulky anion of Congo red are free to move about. The ensuing equilibrium distribution of sodium chloride, governed by the condition of equal activity of any diffusible species on both sides of the membrane, is unequal, tending to prevent chloride (and thus sodium chloride) from entering the Congo red compartment: *Donnan* equilibrium. Unequal electrolyte distribution, in turn, gives rise to a potential difference: *Donnan* potential. Translation of this situation into the exclusion principle of ion exchange membranes is straightforward: The fixed charges of the membrane matrix assume the role of the non-dialysable ions which act to prevent mobile co-ions from entering the matrix: *Donnan* exclusion. Pores or not, a fluid aqueous phase allowing dissolved ions to move within the ion exchange matrix is a logical requirement for this analogy to hold. Electrodialysis, of course, had to wait another thirty years (*Meyer* and *Strauss*, 1940), physiology being the interest of the hour [10].

The only contribution in this essay devoted to membranes proper, also the only multi-authored contribution, is a review by *Bigelow* and *Gemberling* on collodion membranes [11]. First mentioned by *Fick* in 1855 [8], collodion membranes were the first synthetic membranes to compete with natural or processed animal skin like pig's bladder, goldbeater's skin, and parchment paper. Collodion (cellulose nitrate; pyroxylin) is the ancestor of the still thriving family of cellulose ester membranes, which, it may be remembered, generated the first membrane recognized to be asymmetrically structured (*Loeb* and *Sourirajan*, 1960).

Numerous researchers before him must have made the observation which prompted *Kober* to investigate pervaporation, viz., "that a liquid in a collodion bag, which was suspended in the air, evaporated, although the bag was tightly closed" [12]. *Kober's* claim to fame lies not so much in the profoundness of this investigation, but in having named the effect. Ironically, it is his "vacuum perstillation" which today is addressed as pervaporation.

What is missing? Membranes and membrane separations remained laboratory tools until fairly recently, not in the least confined by the fact that there was no polymer research to speak of during the time period documented here. There was little intention of applying membranes to industrial separations, unless the clarification of wine by ultrafiltration using compacted asbestos, introduced towards the end of the 19th century, is considered such. Even notions of medical applications are conspicuously absent, which is all the more surprising considering that the majority of early membrane researchers are biologists, botanists, physiologists or outright medical professionals.

The age of innocence for membrane science ended in 1942. At this time, curiously coinciding, two totally unrelated membrane processes made their appearance, which have changed the world. One is the separation of uranium isotopes by gaseous diffusion of UF_6 (*Manhattan Project*), which, for better or worse, gave access to nuclear energy. The other is hemodialysis (*Kolff*), marking the unsuspecting beginning of the high-tech manipulation of life itself.

Literature cited

[1] J. A. (Abbé) Nollet (1700–1770): Recherches sur les causes du Bouillonnement des Liquides (Investigations on the Causes for the Ebullition of Liquids). Histoire de l'Académie Royale des Sciences, Année MDCCXLVIII, Paris, 1752, 57–104.

[2] R. J. H. Dutrochet (1776–1847): Nouvelles Observations sur l'Endosmose et l'Exosmose, et sur la cause de ce double phénomène (New Observations on Endosmosis and Exosmosis, and on the Cause of this Dual Phenomenon). Annales de Chimie et de Physique, 35 (1827) 393–400.

[3] T. Graham (1805–1869): Notice of the Singular Inflation of a Bladder. Quarterly Journal of Science, No. II (1829) 88–89.

[4] J. K. Mitchell (1793–1858): On the Penetrativeness of Fluids. The Journal of the Royal Institution of Great Britain, No. IV (1831) 101–118; No. V (1831) 307–321.

[5] T. Graham: On the Law of the Diffusion of Gases. The London and Edinburgh Philosophical Magazine and Journal of Science, Vol. II (1833) 175–190; 269–276; 351–358.

[6] M. Knudsen (1871–1949): Die Gesetze der Molekularströmung und der inneren Reibungsströmung der Gase durch Röhren (The Laws of Molecular Flow and of Inner Friction Flow of Gases Through Tubes). Annalen der Physik, 28 (1909) 75–130.

[7] T. Graham: On the Absorption and Dialytic Separation of Gases by Colloid Septa. Part I. Action of a Septum of Caoutchouc. The London, Edinburgh, and Dublin Philosophical Magazine and Journal of Science, Vol. XXXII (1866) 401–420.

[8] A. Fick (1829–1901): Über Diffusion. Poggendorff's Annalen der Physik und Chemie, 94 (1855) 59–86. – Abstracted by the author as: On Liquid Diffusion. The London, Edinburgh, and Dublin Philosophical Magazine and Journal of Science, Vol. X (1855) 30–39.

[9] J. H. van't Hoff (1852–1911): Die Rolle des osmotischen Druckes in der Analogie zwischen Lösungen und Gasen. Zeitschrift für Physikalische Chemie, 1 (1887) 481–508. – Translation published as: The Role of Osmotic Pressure in the Analogy Between Solutions and Gases. In: The Foundations of the Theory of Dilute Solutions, Alembic Club Reprint No. 19, Oliver and Boyd, Edinburgh, 1929, 5–42.

[10] F. G. Donnan (1870–1956): Theorie der Membrangleichgewichte und Membranpotentiale bei Vorhandensein von nicht dialysierenden Elektrolyten. Ein Beitrag zur physikalisch-chemischen Physiologie (Theory of Membrane Equilibria and Membrane Potentials in the Presence of Non-Dialysing Electrolytes. A Contribution to Physical-Chemical Physiology). Zeitschrift für Elektrochemie und angewandte physikalische Chemie, 17 (1911) 572–581.

[11] S. L. Bigelow and A. Gemberling: Collodion Membranes. Journal of the American Chemical Society, 29 (1907) 1576–1589.

[12] P. A. Kober: Pervaporation, Perstillation and Percrystallization. Journal of the American Chemical Society, 39 (1917) 944–948.

Appendix

A   Properties of Aqueous Solutions

Thermodynamic properties of aqueous solutions as function of composition. Upper: NaCl-H$_2$O (nonvolatile solute to saturation); lower: EtOH-H$_2$O (volatile solute completely miscible).

Composition					Osmotic Pressure		Activity Coefficient	
m	x_1	x_2	w %	w %	π_1	π_2	γ_1	γ_2
NaCl	H$_2$O	NaCl	H$_2$O	NaCl	H$_2$O	NaCl	H$_2$O	NaCl
0.01	0.999+	0.001−	99.9+	0.058	0.48	–	–	0.900
0.1	0.998	0.002	99.4	0.58	4.62	–	–	0.767
0.2	0.996	0.004	98.8	1.16	9.16	–	–	0.725
0.5	0.991	0.009	97.2	2.84	22.8	–	–	0.674
1.0	0.982	0.018	94.5	5.52	46.4	–	–	0.660
2.0	0.965	0.035	89.5	10.5	97.5	–	–	0.680
3.0	0.949	0.051	85.1	14.9	155	–	–	0.735
6.0	0.902	0.098	74	26	389	–	–	–
	H$_2$O	EtOH	H$_2$O	EtOH	H$_2$O	EtOH	H$_2$O	EtOH
–	0.99	0.01	97.5	2.5	13.5	1487	1.000	3.028
–	0.95	0.05	88.1	11.9	64.2	863	1.004	2.629
–	0.90	0.10	77.9	22.1	122	632	1.017	2.264
–	0.50	0.50	28.1	71.9	600	210	1.291	1.221
–	0.10	0.90	4.2	95.8	2191	41	2.024	1.009
–	0.05	0.95	2.0	98.0	3030	21.0	2.196	1.002
–	0.01	0.99	0.4	99.6	5140	4.3	2.360	1.000

m = molality of solute [mol/kg]
x = mol fraction (1, solvent; 2, solute)
π = osmotic pressure at 25 °C [bar]
γ_2 for NaCl is the mean activity coefficient at 25 °C

Data compiled and adapted from:

R. W. Stoughton, M. H. Lietzke, Calculation of some thermodynamic properties of sea salt solutions at elevated temperatures from data on NaCl solutions. J. Chem. Eng. Data 10 (1965) 254–260.

A. C. Schneider, C. Pasel, M. Luckas, K. G. Schmidt, J.-D. Herbell, Bestimmung von Ionenaktivitätskoeffizienten in wässrigen Lösungen mit Hilfe ionenselektiver Elektroden. Chem. Ing. Techn. 75 (2003) 244–249.

G. D. Mehta, Comparison of membrane processes with distillation for alcohol-water separation. J. Membrane Sci. 12 (1982) 1–26.

B Criteria of Technical Water Quality

Hardness

Hardness is that part of the total salinity of a water (TDS) associated with the alkaline-earth cations Ca^{++} and Mg^{++}, total hardness comprising all divalent salt ions in solution, while carbonate hardness is the fraction of total hardness which relates to dissolved CO_2. Hardness is expressed as meq/L of hardness-producing species, usually $CaCO_3$ or Ca^{++} (new degrees of hardness according to WHO/EU):

1 meq/L = 0.5 mmol/L = 50 mg/L $CaCO_3$ = 20 mg/L Ca^{++}

A rough classification of water in terms of hardness is

	meq/L	mg/L $CaCO_3$
soft	0–2	0–100
medium hard	2–6	100–300
hard	6–10	300–500
very hard	> 10	> 500

The actual concentration of dissolved Ca^{++} (Mg^{++}) in water depends on the availability of CO_2 which solubilizes $CaCO_3$ ($MgCO_3$) according to

$$CaCO_3 + CO_2 + H_2O \rightleftharpoons Ca(HCO_3)_2 \tag{B.1}$$

Solubility of $CaCO_3$ in pure water is little more than 10 mg/L, while a concentration of CO_2 corresponding to its ambient partial pressure yields a carbonate hardness of about 100 mg/L (2 meq/L). Scale inhibitors like polymeric phosphates, which are widely used in water desalination, will sustain a concentration of up to 250 mg/L $CaCO_3$ (5 meq/L). WHO recommendation for drinking water is 2 meq/L as highest desirable level, and a maximum permissible total hardness of

10 meq/L. The EU drinking water directive suggests a minimum hardness for softened (demineralized) water of 3 meq/L.

Carbonate hardness may be reduced through addition of lime to precipitate $CaCO_3$ (lime softening), according to

$$Ca(HCO_3)_2 + Ca(OH)_2 \rightarrow 2\,CaCO_3 + 2\,H_2O \tag{B.2}$$

Lime softening is a process of partial demineralization used extensively in the treatment of brackish groundwaters. Overall reduction of TDS is of the order of 10%, depending on the fraction of divalent ions in solution. Conversely, hardness is conveyed to soft waters by adding lime in combination with CO_2,

$$Ca(OH)_2 + 2\,CO_2 \rightarrow Ca(HCO_3)_2 \tag{B.3}$$

This is a method of posttreatment used to remineralize the product water of reverse osmosis or thermal seawater desalination which, on account of its negligible salinity, is extremely soft and therefore aggressive (see below).

Alkalinity

The ratios of the carbonate-containing species in aqueous solution are a function of pH value (and vice versa), as illustrated by the following figures:

	CO_2 / HCO_3^-	$HCO_3^- / CO_3^=$
pH 6	2	20000
pH 7	0.2	2000
pH 8	0.02	200

The standard acid consumption on titration of a water sample to pH 4.3 is the total alkalinity (acid capacity) of the water, K_A 4.3, reported as mmol/L acid (HCl or HCO_3^-) or as mg/L $CaCO_3$. At pH 4.3 the concentration of bicarbonate (HCO_3^-) is down to 1% of the dissolved CO_2, and neutralization of carbonate ($CO_3^=$) is essentially complete. Alkalinity is thus related to carbonate hardness, specifically to HCO_3^- concentration, and is a measure of the capacity of the water to resist changes in pH (buffering). A minimum alkalinity of 0.5 mmol/L HCO_3^- is suggested by the EU drinking water directive.

Alkalinity (carbonate hardness) is destroyed by addition of acid, with simultaneous generation of CO_2,

$$Ca(HCO_3)_2 + 2\,HCl \rightarrow CaCl_2 + H_2O + 2\,CO_2 \tag{B.4}$$

Acidification followed by aeration to remove excess CO_2 is a common pretreatment practice in water desalination to reduce hardness and to prevent carbonate scales from being deposited. In thermal seawater desalination, scale formation proceeds by the reverse of reaction B.1 at elevated temperature, again producing CO_2.

Corrosiveness

Any excess of CO_2 beyond that needed to keep Ca^{++} (Mg^{++}) in solution according to Eq. B.1 is termed aggressive and must be removed before the water is distributed. Excess CO_2 is harmful in two ways: Firstly, dissolved excess CO_2 acts much like a mineral acid, particularly in the presence of oxygen, and will corrode the ferrous ducts of the distribution system; secondly, excess CO_2 interferes with the corrosion-protective mineral layer in ferrous ducts, either by preventing its formation or by re-dissolving it. Formation of this protective layer, which basically consists of crystalline $CaCO_3$ and $FeCO_3$, requires a minimum alkalinity to exclude free CO_2, a residual hardness of at least 30 mg/L $CaCO_3$, and dissolved oxygen.

Very soft waters, even after thorough aeration (deacidification), always retain some excess CO_2 (typically 5 mg/L) and thus are aggressive on both above counts. Only with very hard waters a small excess of CO_2 may be tolerated, aeration then bearing the risk of precipitation of $CaCO_3$ according to Eq. B.1. – Mixing of waters of different carbonate hardness always yields aggressive water requiring deacidification.

Yet another kind of corrosion is due to sulfate ions ($SO_4^=$) decomposing cement materials (concrete). Again, soft waters are more aggressive than hard water.

The above interrelations are especially significant for desalinated water supplies. Water produced by desalination invariably is soft and acidic, regardless of the process used: The product water of thermal desalination is essentially devoid of all saline constituents, and is likely to contain free CO_2 originating from the acid pretreatment,

Eq. B.4. In reverse osmosis desalination there is a preferential rejection of the hardness-producing ions by the membranes, resulting in an enrichment of CO_2 (lowering of pH) and a relative increase of monovalent (soft) ions in the permeate.

Corrosion control is through chemical treatment and appropriate material selection. The European directive addressing posttreatment of demineralized drinking water recommends removal of excess CO_2, raising the pH value to 8, and remineralization to a comparatively high 3 meq/L (corresponding to 150 mg/L $CaCO_3$). According to an industrial standard, minimum remineralization for bulk transport of desalinated seawater is 35 mg/L $CaCO_3$.

C Marker Molecules

Membrane filtration: Molecular mass of frequently encountered nonelectrolytes and macromolecules (marker molecules). This is a teaching aid. Entries are compiled from various open literature sources (including membrane manufacturer's brochures), and are not critically weighted. Molecular mass > 5000 usually is quoted as "average", individual assay depending on origin (biological matrix) and method of mass determination.

D-Alanine	89	amino acid
Creatinine	113	(urine constituent)
Phenylalanine	165	amino acid
Glucose	180	blood sugar
Tryptophan	204	amino acid
Sucrose	342	cane/beet sugar
Lactose	342	milk sugar
Raffinose	504	tri-saccharide
Vitamin B12	1355	(in activated sludge)
Bacitracin	1400	antibiotic polypeptide (globular)
Inulin	5200	polysaccharide
Insulin	5800	polypeptide hormone
β2-Microglobulin	11800	plasma protein
Cytochrome C	13000	respiratory proteid (globular)
Lysozyme	14400	mucolytic enzyme
α-Lactalbumin	16000	cheese whey protein
Myoglobin	17500	respiratory proteid
ß-Lactoglobulin A	18700	milk protein
Trypsin	24000	proteolytic enzyme
Chymotrypsinogen A	24500	proteolytic enzyme
Carbonic anhydrase	31000	CO_2 hydrating enzyme

Pepsin	34500	gastric enzyme (globular)
Ovalbumin	45000	egg white protein
Bovine serum albumin (BSA)	67000	plasma protein (globular)
Hemoglobin	68000	respiratory proteid
Human serum albumin	69000	plasma protein (globular)
Transferrin (s)	80000	iron-binding glycoprotein(s)
Phosphorylase B	94000	plasma enzyme
Aldolase	142000	plasma enzyme (lyase)
Immunoglobulin IgG	160000	antibody protein (globular)
Catalase	240000	anti-peroxide enzyme
Ferritin (apoferritin)	450000	iron storage protein
Myosin	500000	muscle protein
Thyroglobulin	680000	gluco protein (thyroid)
Immunoglobulin IgM	960000	antibody protein

Synthetic macromolecules

Polyethylene glycol (PEG)	200 to	2×10^6
Polyvinylpyrrolidone (PVP)	2500 to	900000
Dextran (polysaccharide)	15000 to	50×10^6

D   Membrane Polymers

A survey of organic membrane polymers, arranged in order of approximate glass transition temperature (°C). It is noted that by the complex nature of the polymeric state of matter (chemical and physical structure) the temperature of glass transition does not have the quality of a natural constant. Accordingly, published figures vary; the data presented are meant to provide orientation. Hydrophilic polymers are set in *italics*.

Natural polymeric materials

Natural rubber (polyisoprene)		– 70
Cellulose (regenerated)	*CE*	
Cellulose derivatives (polycellobiose)		
Ethylcellulose (ether)	*EC*	+ 45
Cellulose nitrate (ester)	*CN*	+ 60
Cellulose diacetate	*CA*	+ 70
Cellulose triacetate	*CTA*	+ 100

Synthetic polymers

Polydimethylsiloxane	PDMS	– 120
Polybutadiene	PB	– 80
Polyetherblockamide	PEBA	– 65
Polyethyleneoxide	*PEO*	– 50
Polyethylene (40 % cryst.)	LDPE	– 70
Polyethylene (70 % cryst.)	HDPE	– 20
Polyvinylidenefluoride	*PVDF*	– 40
Polypropylene	PP	– 15
Polyvinylacetate	PVAC	+ 30

Polyamide (Nylon)	*PA*	*< 100*
Polyvinylalcohol	*PVAL*	*+ 85*
Polyvinylchloride	PVC	+ 90
Polystyrene	PS	+ 100
Polymethylmethacrylate	*PMMA*	*+ 110*
Polyacrylonitrile	*PAN*	*+ 120*
Polytetrafluoroethylene (Teflon)	PTFE	+ 125
Polyetheretherketone	PEEK	+ 140
Polycarbonate	*PC*	*+ 150*
Polyphenyleneoxide	*PPO*	*+ 170*
Polyvinylpyrrolidone (360 000)	*PVP*	*+ 180*
Polysulfone	PSU	+ 190
Polytrimethylsilylpropyne	PTMSP	> 200 (?)
Polyphenylsulfone	PPS	+ 215
Polyetherimide	PEI	+ 215
Polyethersulfone	*PES*	*+ 230*
Polyamide (aromatic)	*PA*	*+ 270*
Polyimide	PI	> 300 (?)
Polyethylene terephthalate	PET	(mp. ~ 240)

Synthetic copolymers (elastomers)

Acrylonitrile-butadiene	NBR
Acrylonitrile-butadiene-styrene	ABS
Styrene-butadiene	SBR
Ethene-propene-diene	EPDM
Ethylene-vinylalcohol	EVAL

E Microporous Structures

A collection of electron micrographs (SEM = scanning electron microscopy), illustrating the scope and variety of organic and inorganic porous materials, along with some indication of the imaginative methods by which porous barriers are created.

Fig. E.1 Kieselguhr: Deposits of the silica shells of unicellular algae (diatoms) from the Tertiary geologic period. When compacted into cylinders: *Berkefeld* filter for drinking water disinfection. (When employed as absorbant for nitroglycerin: *Dynamite*). – [x 1200; Meyer-Breloh].

Fig. E.2 Glass fiber depth filter. Nonwoven (randomly compacted) filters from fibers are ancient clarifying aids, their current rendition being polymeric nanofiber devices. There is an anticipated relationship between fiber diameter and retention capability: Asbestos, 20 nm (outlawed); glass fibers, thinness limited by health hazard considerations; polymeric nanofibers (by electrospinning), 100 to 1000 nm (0.1 to 1 µm); stainless steel, 4 µm; human hair, 70 µm. – [x 1000; FZK].

Fig. E.3 Cellulose acetate (CA): A member of the generic class of cellulose derivatives, capable of forming homogeneous as well as porous barriers. Shown is the porous substructure of an integral-asymmetric ("skinned") reverse osmosis membrane produced by solvent-nonsolvent phase inversion according to Loeb-Sourirajan. – [GKSS].

Fig. E.4 Polysulfone (PSU), the workhorse of ultrafiltration and widely used as support for composite membranes. As generic class, which includes sulfonated polysulfones as well as polyethersulfone (PES), polysulfone is versatile: Itself hydrophobic, hydrophilicity is imparted through sulfonation. Discernible is a structure of microscopic nodules and nodule aggregates. – [Rated 0.8 µm; Seitz].

Fig. E.5 Polyamide (aliphatic PA, Nylon 66). An interlaced structure produced by a proprietary foaming process. Nylon is a trade name for a series of related polymers containing the –CONH– bond in its structure. It is also the "hard" component in segment-elastomeric block-copolymers like PEBA. – [Rated 0.2 µm; x 3000; PALL].

Fig. E.6 Polyamide (aromatic PA). The original, and still going, "FT-30" membrane for seawater desalination by reverse osmosis. In view are surface and fractured edge of the composite membrane produced by interfacial polymerization onto a polysulfone porous support (Fig. E.4). Note the highly corrugated "skin" of this barrier. – [FilmTec/Dow; Chapter 3, Ref. (4)].

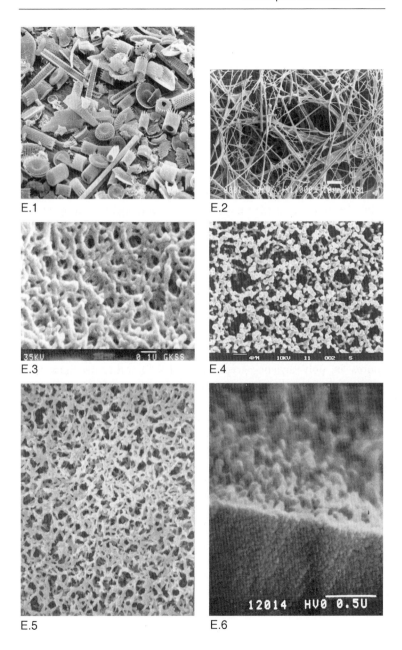

E.1

E.2

E.3

E.4

E.5

E.6

Fig. E.7 Polypropylene (PP). A structure of interconnected (open) vesicles produced by thermal inversion from a heated polymer solution. Porosity (vesicular size) is adjustable by controlling the rate of cooling. The picture shown is a section out of a tubular membrane of gradually narrowing (unisotropic) pore structure [Akzo].

In a related process, precipitation is from a polymer-diluent system initially homogenized by heating to above the melting temperature of the polymer; PP in mineral oil [TIPS = thermally induced phase separation (Lloyd)].

Fig. E.8 Polymer blend: Polyethersulfon (PES)/polyamide (PA) blended with polyvinylpyrrolidone (PVP). An asymmetric (unisotropic) globular structure with randomly distributed hydrophilic microdomains in a hydrophobic matrix (for use in hemodialysis). – [Rated 6 nm; Gambro].

Fig. E.9 Track-etched pores: Surface of a capillary pore membrane (complete with retained asbestos fibers). Uniform capillary pores are created by irradiating melt-extruded polymer film followed by chemical etching of the nucleation tracks; applied to polymers which are intractable by solution casting: Polycarbonate (PC) and polyester (PE). – [Nuclepore].

Fig. E.10 Cross section of a track-etched pore structure: Capillary pores in polyethylene terephthalate (PET), otherwise known as a fiber forming polyester. – [Pore diameter 0.7 µm; GSI].

Fig. E.11 Pores by controlled stretching of semicrystalline polymer film. Slit-shaped pores (ruptures) in melt-extruded polypropylene (PP). – [Slit rating 0.02 by 0.2 µm; Celanese/Celgard].

Fig. E.12 A tortuous pore structure by expansion (biaxial stretching) of polymer film, Teflon (PTFE): Nodules interconnected by fibrils suggestive of the direction of stretching. – [Gore/Gore-Tex].

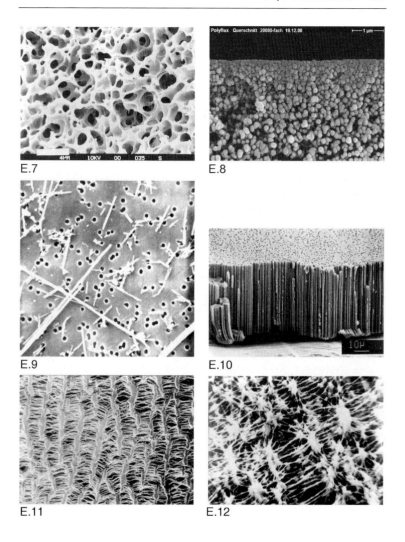

E.7

E.8

E.9

E.10

E.11

E.12

Fig. E.13 Alumina (aluminum oxide) barrier: An asymmetric inorganic membrane composed of microporous layers on a macroporous mineral support, produced by slip-coating and sintering reminiscent of ceramic techniques. – [Rated 0.2 μm; Alcoa/Ceraver].

Fig. E.14 Alumina (aluminum oxide): A straight channel structure obtained by anodic oxidation of metallic aluminum, channel width 0.2 μm. By controlling the voltage the channels divide near the surface to yield a 0.02 μm (20 nm) asymmetric pore structure as shown [Alcan/Anotec].

Fig. E.15 Porous glass (CPG, controlled pore glass). Obtained by a process of micro-dispersed phase separation followed by leaching of the dispersed phase. A structure of interconnected nodules reminiscent of porous polysulfone. – [Rated 85 nm; Schott].

Fig. E.16 Pure metallic silver, formed from suspensions of amorphous silver into molecularly bonded (that is, not sintered) microporous membranes of 50 μm thickness; antiseptic and reusable. – [Graded 0.2 to 5 μm; Osmonics].

Fig. E.17 Asymmetric metallic barrier: A filter mat of randomly compacted (nonwoven) stainless steel fibers (see E.2) sintered onto a support of porous stainless steel made by powder metallurgy. Fiber thickness, 4 μm; active layer, 200 μm; total thickness, 3 mm. – [Rated 1 μm; Krebsöge].

Fig. E.18 Bacterial S-layers. The crystalline protein envelopes of certain archeo-bacteria are isolated and deposited on a filter support to form highly ordered two-dimensional grids with application potential as isoporous nanofilters. Uniform size of the underlying bacteria makes for uniform pore size (2 to 6 nm depending on spcies) and "sharp" molecular weight cutoff (near MW 50000), the "membranes" having the solvent stability of the archeo-proteins. Shown is a computer image reconstruction of an hexagonal grid. – [Nanosearch/Biofil].

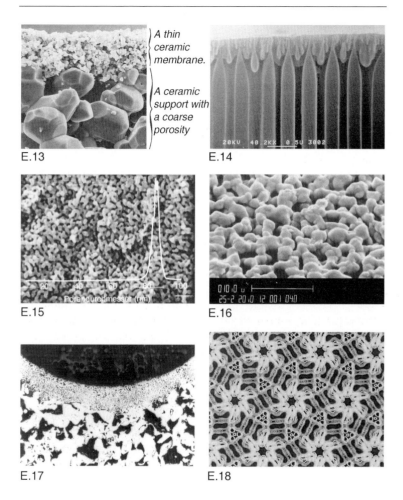

A thin
ceramic
membrane.

A ceramic
support with
a coarse
porosity

E.13

E.14

E.15

E.16

E.17

E.18

Subject Index

Printing: Krips bv, Meppel, The Netherlands
Binding: Stürtz, Würzburg, Germany